MATTER
AND LIGHT
The New Physics

BY

LOUIS DE BROGLIE

Membre de l'Institut
Nobel Prize Award, 1927
Professeur à la Faculté
des Sciences de Paris

TRANSLATED BY W. H. JOHNSTON, B.A.

New York · W · W · NORTON & CO · INC · *Publishers*

FIRST EDITION PUBLISHED IN ENGLISH IN 1939
BY W · W · NORTON & COMPANY INC
70, FIFTH AVENUE, NEW YORK

PRINTED IN GREAT BRITAIN

TRANSLATOR'S NOTE

THE Author has in certain places modified the original French text for the English translation, for the sake of greater cohesion, and has also revised some passages, in order to bring them into accord with the results of later research. Occasional Translator's Notes are shown in square brackets.

The chapter on "The Undulatory Aspects of the Electron" has the special historical interest of having been delivered as a Lecture on the occasion of the Author's receipt of the Nobel Award, while that on "Wave Mechanics and its Interpretations" was given as an Address at the Glasgow meeting of the British Association in 1928.

I am indebted to Dr. J. E. Turner, of the University of Liverpool, for assistance with the translation and the proofs, and to Dr. C. Strachan, of the same University, I am indebted for valuable assistance in dealing with the equations and the more technical passages, as well as for reading the proofs.

W. H. J.

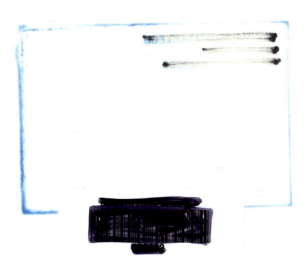

PREFACE

THE amiable insistence of my friend André George has induced me to collect in the present Volume a number of Studies on contemporary Physics written from both the general and the more metaphysical point of view. Each of these Studies forms an independent whole, and can be read by itself. A slight degree of repetition—which the reader is asked to overlook—has been the inevitable result: for on more than one occasion I have been compelled to duplicate a summary of the great fundamental stages of contemporary Physics, such as the classification of simple substances, the investigation of the photo-electric effect and the origin of the Theory of Light Quanta and of Wave Mechanics: the subjects are somewhat technical, and I cannot well assume that they are common knowledge. But though the same subject is outlined in several of these Studies, I have tried to take up a different point of view in each, and have endeavoured to throw light on different aspects of the essential problems of Quantum Physics in order to facilitate a grasp of their importance.

On comparing the different chapters the reader will observe that, while overlapping, they also complement one another; and he will feel the fascination and greatness inherent in the vast structure of modern Physics. And while admiring the vast number and the extreme delicacy of experimental facts which laboratory physicists have succeeded in revealing, and the strange and brilliant concepts devised by theorists to explain them, he will appreciate to what a degree the methods and ideas of physicists have grown in subtlety during recent years, and how great has been the progress from the somewhat ingenuous Realism and the over-simplified Mechanics of earlier thinkers. The more deeply we descend into the minutest structures of Matter, the more clearly we see that the concepts evolved by the mind in the course of everyday experience —especially those of Time and Space—must fail us in an endeavour to describe the new worlds which we are entering. One feels tempted to say that the outlines of our concepts must undergo a progressive blurring, in order that they may retain some semblance of relevance to the realities of the subatomic scales. Time and Space, in other words, are too loose a dress for the elementary entities; individuality becomes attenuated in the mysterious processes of interaction, and even Determinism, the darling of an older generation of physicists, is forced to yield. But the great book of Science is never finished: other surprises await us: who knows what mysteries are hidden within the nucleus of an atom, which, although a million million times smaller than the smallest living thing, is yet a universe in itself?

Thus a few decades have witnessed the downfall of the best established principles and most firmly supported conclusions: a fact which shows the need of caution in basing general metaphysical principles on the advances of the different sciences. Such a process would be tantamount to building on ever-shifting foundations; and once we have grasped by how much the sum of what we do not know exceeds the sum of what we do, we shall feel little taste for precipitate conclusions. Yet it is fair to observe that the advance made by Quantum Physics has opened entirely novel perspectives on a great number of questions, and that the future orientation of metaphysical doctrines will almost inevitably be deeply influenced sooner or later. And it is equally fair to observe in passing that no less a physicist than Niels Bohr thinks that the "uncertainties" and "complementary aspects" of Quantum Physics are sure to find a place sooner or later in biological theory. For according to Genetics, all the essential factors of life and heredity are contained in elements so minute as to be more or less comparable with atoms; it is possible even that they are contained in fractions of these elements: so that Bohr's suggestion becomes all the less surprising, since, if he is right, the mysterious interconnections of Life and Matter would take place in so extremely restricted a sphere that Quantum concepts would necessarily be operating. A detailed study of questions of this order would, however, be certainly premature, and I shall leave the task of meditating on them to the philosophers among my readers.

In concluding these observations I wish to say a few words on a question which has at all times engaged scientific thinkers. What is the value of Science? in other words, what are the reasons which induce us to devote ourselves to, and to admire, scientific investigation? Many people value it in proportion to its practical applications. They point to the many material improvements in our everyday existence due to it, and to the powerful instruments it has given us for preserving and to some extent prolonging our life; and they voice the hope that Science will open to us a prospect of practically infinite progress. But such a view certainly cannot stand without reservation. For not all the applications of Science are beneficent, nor is it clear that its future course will ensure a real advance to humanity, since it is quite certain that such progress depends much more on man's intellectual and moral level than on the material conditions under which he lives. Yet it must be admitted that applied Science has rendered certain aspects of our everyday existence more pleasant and gracious, and that it can continue this wholesome work if we have the wit to merit it. We are thus fully entitled to admire Science for its applied value, and for the comforts and material goods by which it has enriched human life; though we must not forget that by its very nature man's life must always remain uncertain and wretched. And I think that another reason can be found for admiring scientific effort: we can learn to admire it by appreciating the value of that for which it stands. For it is with Science as it is with all the greatest values in life: it is only on the spiritual plane that its full stature is attained. Science exacts our admiration because it is one of the great works of the human spirit.

With two exceptions the following chapters require no Mathematics. The exceptions consist in the chapter on "The Present State of Electromagnetic Theory" and that on "The Undulatory Aspects of the Electron." They can be omitted without prejudice to an understanding of the rest of the book.

CONTENTS

I

A GENERAL SURVEY
OF
PRESENT-DAY PHYSICS

I

THE PROGRESS OF CONTEMPORARY PHYSICS

By way of an Introduction to the series of Studies in the present Volume, I wish to deal with the progress of contemporary Physics from a comprehensive viewpoint. It is a very wide subject, for progress has been made in so many different directions, and at so swift a rate, that it would require many books to set them out in detail. Further, the subject is one whose treatment is difficult, since Physics has become a science of growing complexity, tending towards higher and higher levels; so that it is far from easy to explain the progress it has made with adequate exactness and precision without at the same time employing rather complicated ideas and methods of reasoning.

But though difficult, the subject is still well worth discussing. Its study leads to a comforting and encouraging impression: it suggests that the human spirit, despite the difficulties implied in the very conditions of material life, is following triumphantly a long, and an ascending, course. Man's endeavours to know, and to understand, more, are like physical training: for as the latter gives suppleness to the athlete's limbs and prepares him for victories to come, so the former render an analogous service to the mind, and adapt it to the morrow's progress. In this way the volume of our knowledge and the subtlety of our concepts have slowly increased in the course of time, each generation building on the results obtained by its predecessors, and each result being in its turn surpassed—although without it no generation of men could have made any advance at all.

The claim that the rate of this evolution has increased at the present time would seem to be indisputable—a fact to provide

legitimate reassurance and encouragement amidst the disturbances
and shocks so common in modern life. And nowhere else, perhaps,
is a more powerful impression of continued progress and renova-
tion to be had than from a study of the wonderful advance of
Physics during the last forty years.

Like all the other natural sciences, Physics advances by two
distinct roads. On the one hand it operates empirically, and thus
is enabled to discover and analyse a growing number of phenomena
—in this instance, of physical facts; on the other hand it also operates
by theory, which allows it to collect and assemble the known
facts in one consistent system, and to predict new ones for the
guidance of experimental research.   In this way the joint efforts
of experiment and theory, at any given time, provide the body of
knowledge which is the sum total of the Physics of the day.

At the beginning of the development of modern Science, it was
naturally enough the study of the physical phenomena which we
observe immediately around us that first drew the attention of
physicists. Thus the investigation of the equilibrium and the motion
of bodies led to the development of the branch of Physics—today
an independent study—known as mechanics. Similarly research
into the phenomena of sound led to acoustics, while optics was
created by collecting the phenomena of light and forming them
into one system.

The great task and the splendid achievement of nineteenth-
century Physics consisted in thus increasing the exactness and
range—in every direction—of our knowledge of the phenomena
taking place on the human scale. Not only did it continue to
develop mechanics, acoustics and optics—the leading branches of
classical Science—but it also created on every side new sciences
possessing innumerable aspects, such as thermodynamics and the
science of electricity.

The mastery of the vast sphere of facts covered by these various
branches of Physics has enabled both abstract students and technical workers to draw thence a great number of practical applications. The inventions—ranging from the steam engine to wireless
telephony—derived from the nineteenth-century advance of

Physics, the benefits of which we enjoy today, are innumerable; and these inventions play so important a part, directly or indirectly, in the everyday life of each of us that it would be wholly superfluous to enumerate them.

In this way, then, nineteenth-century Physics succeeded in achieving the complete domination of the phenomena we observe around us. No doubt research into these phenomena can still lead to the knowledge of many further facts and to new applications; yet it appears that in this sphere the essential work has now been completed. And, in fact, during the last thirty or forty years the attention of pioneers in Physics has been turning increasingly towards more subtle phenomena, which could be neither discovered nor analysed without an extremely refined experimental technique: molecular, atomic and intra-atomic phenomena. The fact is that in order to satisfy human curiosity it is not enough to know the behaviour of material bodies taken as wholes, or in their manifestations *en masse*, or to grasp the reactions between Light and Matter when observed on the macroscopic scale: what is required is to descend to individual details, to attempt the analysis of the structure of both Matter and Light, and to specify the elementary processes which in their totality constitute the macroscopic phenomena. It is a difficult inquiry, and for its success an extremely delicate experimental technique is required, capable of discovering and recording exceedingly subtle events, and of measuring exactly magnitudes vastly smaller than those occurring in our everyday experience. Still further, bold theories are required, based on the highest branches of Mathematics and prepared to make use of entirely novel similes and concepts. Hence we can infer the amount of ingenuity, patience and talent needed for the formulation and advancement of this atomic Physics.

*   *   *

On the experimental side, then, the progress made has been characterized by a daily growing knowledge of the ultimate constituent entities of Matter and of the phenomena connected with the existence of these ultimate constituent entities.

Chemistry had long assumed that material substances are composed of atoms; and the actual investigation of the properties of material substances shows them to be divided into two classes: compound substances, which can be reduced to simpler ones by appropriate methods; and the simple substances themselves—the chemical elements—which resist any attempt at such reduction. In the next place, the study of the quantitative laws, in accordance with which the simple substances combine to form compounds, led chemists during last century to adopt the following hypothesis:

"A simple substance is supposed to be formed of small particles, all identical with each other, called the atoms of this element; compounds, on the other hand, are supposed to be formed of molecules resulting from the combination of a number of the atoms constituting the simple substances." According to this hypothesis, therefore, a composite substance is broken up by reducing it to the elements of which it is composed, which means that its molecules are disintegrated and the atoms which they contain set free. The number of these simple substances known today is 89, but it is believed that their total number is 92 (or possibly 93). All material substances, therefore, are regarded as constructed from 92 different kinds of atoms.

The Atomic Theory not only succeeded in introducing order into Chemistry: it also extended into the domain of Physics. For if material substances are composed of molecules and atoms, then their physical properties must be capable of explanation in terms of their atomic structure. The properties of the various gases, for example, must be explicable on the assumption that a given gas consists of an immense number of molecules or atoms in rapid motion; the pressure of a gas on the wall of the containing vessel will then be due to the impacts of the molecules against the wall, while the temperature of the gas will be the measure of the average of the motions of the molecules, which increase as the temperature rises. During the second half of the nineteenth century, this view of the structure of gases was developed under the name of the Kinetic Theory of Gases, and it enables us to understand the origin of the laws governing the behaviour of gases as discovered experimentally. For if the Atomic Theory is correct, then the properties of solids and liquids must be capable of interpretation on the assumption that, in the solid and the liquid states, the molecules or atoms are much closer to each other than in the gaseous state. Thus there is an interplay of considerable forces between atoms and molecules in these states, and these should account for such characteristic properties of solids and liquids as incompressibility and cohesion. The Atomic Theory of Matter, again, has been confirmed by brilliant direct experiments such as those of Jean Perrin, by means of which it has been possible to measure the weights of different kinds of atoms and to find their number per cubic centimetre.

Without entering further into the evolution of the Atomic Theory I shall confine myself to recalling that in Physics, just as in Chemistry, the theory which assumed that all substances consist of molecules, which in turn consist of different combinations of elementary atoms, proved very fruitful in practice and can hence be fairly regarded as a useful statement of the actual facts. But physicists did not rest content at this point. They wished further to discover the structure of the atoms themselves, and to understand the *differentiae* subsisting between the atoms of the different elements; and in this research they were aided by our increasing knowledge of electrical phenomena. When these phenomena first began to be investigated it appeared expedient to treat, for example, the electric current passing through a metallic wire as though it were tantamount to the passing of an "electric fluid" through the wire. But we know that there are two kinds of electricity—positive and negative. Hence it is natural to assume that there are two fluids: the positive and the negative electrical fluid. These fluids, again, can be imagined in two different ways: we may imagine that they consist of a substance uniformly occupying the whole of the space where the fluid is; or we may imagine that they consist of clouds of little corpuscles each of which is a minute sphere of electricity. Experiment, however, has decided in favour of the second view, and some thirty years ago it showed that negative electricity consists of minute corpuscles which are all identical, and have a mass and an electric charge of extremely small dimensions, called electrons. These have been successfully segregated from Matter in bulk, and their behaviour when moving in empty space has been observed; and it has been found that in fact they move in the way in which small particles, electrically charged, ought to move in accordance with the Laws of classical Mechanics; while by observing their behaviour in the presence of electrical or magnetic fields it has proved possible to measure both their charge and their mass—which, I repeat, are extremely small. The demonstration of the corpuscular structure of positive electricity, on the other hand, is less direct; nevertheless physicists have come to the conclusion that positive electricity, too, is subdivided into corpuscles which are identical with each other, today known as protons.[1]

The proton has a mass which, though still extremely small, is nearly 2,000 times greater than that of the electron, a fact indicative of a curious asymmetry between positive and negative electricity. The charge of the proton, on the other hand, is equal to that of the electron in absolute value, but of course bears an opposite sign, being positive and not negative.

Electrons and protons, then, have extremely small mass. This mass, however, is not equal to zero, and a really vast number of protons and electrons may make up a fairly considerable total mass. Hence it is tempting to assume that all material substances—whose essential characteristics consist in the fact that they possess weight and inertia, in other words, that they have mass—consist in the last analysis exclusively of vast numbers of protons and electrons. On this view the atoms of the elements, which are the ultimate fabric of which material substances are composed, should themselves consist of electrons and protons; and the 92 kinds of different atoms, already referred to, of which the 92 elements are composed, should be 92 different combinations of electrons and protons. The idea that atoms consist of protons and electrons was next formulated in more exact terms as the result of the experiments of the great British physicist, Lord Rutherford, and of the theoretic work of the Danish scientist, Niels Bohr. The atom of a simple

[1] But. cf. further pp. 75 ff.

substance was thus shown to consist of a central nucleus, having a positive charge equal to a whole number N times as great as the charge of the proton, and of N electrons gravitating around the nucleus. The entire system, therefore, is electrically neutral, and the nucleus itself is doubtless formed of protons and electrons in the way which we shall see in greater detail below.[1] Almost the entire mass of the atom is concentrated in the nucleus, for the latter contains protons, and these in turn are very much heavier than electrons. The Hydrogen atom is the simplest of the atoms, and consists of a nucleus formed by a single proton around which a single electron revolves. The atom of one element is differentiated from that of another by the number N of positive elementary charges which the nucleus carries. Simple substances can thus be arranged in a series according to the ascending value of the number N, beginning with Hydrogen (N = 1) and ending with Uranium (N = 92). It has been found that this way of classifying substances agrees with that which had been inferred from the value of their atomic weights and from their chemical properties, an arrangement known as Mendelejeff's classification, after the name of the Russian chemist who first proposed it.

I cannot here explain in detail why the idea that the atom is a kind of miniature solar system, with the nucleus for sun and the electrons for planets, has met with so much favour from physicists. I will only say that it has provided an interpretation, not only of the chemical properties of simple substances, but also of several of their physical properties, such as the light rays which they can emit in certain circumstances, for example when incandescent.

One point, however, must be noted. In order to achieve a satisfactory formulation of the theory that the atom is equivalent to a kind of solar system, Bohr had to import a foreign idea, borrowed from the Quantum Theory previously worked out by Planck. I said above that in the experiments in which we are able to follow the motion of an electron, the latter behaves like a small corpuscle of very slight mass, and that its motion can be predicted by applying the Laws of classical Mechanics. Let us consider, however, the

[1] cf. pp. 75 ff.

motion of an electron along a particularly short trajectory. We cannot follow this motion by actual observation; but Bohr has done so in imagination, in order to calculate the characteristic properties of the atom when treated, as he treats it, as a planetary system. Planck, indeed, was himself the first to find that this motion cannot conform exactly to the laws of classical Mechanics. For among the totality of movements which classical Mechanics regards as possible, those which the electron can in fact execute form only a fraction: and this latter privileged group have been called "quantized." Bohr therefore, in his theory of the parallelism between the atom and the solar system, has been forced to incorporate Planck's idea and has found that, in fact, the planet-electrons can only have quantized motion; and it is this fact which, in a measure, has provided the key to all the properties of the atoms.

Let us now sum up. Investigation of the properties of material substances has led physicists to treat Matter as consisting solely of small corpuscles, called electrons and protons.[1] Various combinations of these corpuscles constitute the atoms of the 92 simple substances which form the raw material of the molecules from which compounds are built up. Such was the conclusion reached some 20 years ago; but we shall shortly see that conditions have since become far less simple; for the moment, however, we must leave the subject of Matter and turn to that of Light.

* * *

When Light reaches us from the sun or the stars it comes to the eye after a journey across vast spaces void of Matter. It follows from this that Light can cross empty space without difficulty, wherein it differs for example from sound, since it is not bound up with any motion of Matter. Hence a description of the physical world would remain incomplete unless we were to add to Matter another reality independent of it. This entity is Light.

Now what is Light? What is its structure?

The ancient philosophers, and many scientists until the beginning of last century, maintained that Light consisted of minute corpuscles in a state of rapid motion; and the fact that Light travels in straight

[1] cf. previous Note.

lines under ordinary conditions, and its reflection in a mirror, are explained at once by this hypothesis.

But the corpuscular Theory of Light was abandoned entirely, about a century ago, in consequence of the work of the English physicist Young and, even more, of the research of a brilliant French scientist, Augustin Fresnel. Actually, Young and Fresnel discovered a whole set of luminous phenomena—those of interference and of diffraction—which could not be accounted for at all on the corpuscular Theory, while the adoption of another concept—the Wave Theory of Light—accounts both for the classical phenomena of motion in a straight line, of reflection and refraction, and also for the phenomena of interference and diffraction. Fresnel's demonstration of all this was an admirable one.

The Wave Theory of Light—which had previously been adopted by the Dutch scientist Christian Huyghens and other far-sighted thinkers—holds that the propagation of Light should be compared to that of a wave in an elastic medium, like the ripples which travel on the surface of a sheet of water when a stone is thrown in. And since Light moves in empty space, Fresnel assumed the existence of a particularly subtle medium—the Ether—supposed to penetrate all material substances, to fill empty space and to act as vehicle for the light-waves.

Let me now explain the way in which a wave is to be imagined. When a wave moves freely it may be compared to a succession of ripples in water, their crests being separated by a constant distance known as the wave-length. The entire group of these ripples moves in the direction of propagation with a certain velocity:—that at which the wave advances. For light-waves in empty space this velocity has been shown by experiments, made after Fresnel's death, to be 300,000 kilometres per second.[1] The different waves with their crests and troughs pass a given point in space in succession; and at this point, whatever magnitude it is that is travelling in the form of waves must pass through a periodic variation, the period itself being obviously equal to the time elapsing between the passing of two consecutive crests.

[1] More precisely, 299,764± 15 km. per sec. *Science Progress*, XXXII, 716.

The three magnitudes—the velocity, length and frequency of the wave—are not independent of each other, the frequency being obviously equal to the velocity divided by the wave-length.

We have seen how a wave advances in a region where there is nothing to interfere with its propagation. But conditions are different when the wave meets with an obstacle in the course of its journey; for example if it meets with a surface which stops or reflects it, or again if it has to pass through an aperture in a screen, or if it meets particles of matter which diffract it. In such a case the wave will be deformed and turned back on itself, with the result that instead of a simple wave we shall have a multiplicity of simple, but superimposed, waves; and then the resulting type of vibration, at any given point, depends on the way in which the simple superimposed waves tend to reinforce or to enfeeble each other. If there is an additive effect as between the various simple waves, or if they are in phase, as it is called, then the resulting vibration will be one of great intensity; while if their phases are in opposition, the resulting vibration will be weak or even non-existent. To sum up, the existence of obstacles interfering with the propagation of a wave brings about a complicated distribution of the various intensities of vibration, the distribution depending in the main on the wave-length of the wave which meets with the obstacle in question. Of this type are the phenomena of interference and diffraction.

If now we adopt the idea that Light consists of waves, we are led to expect that if there is an obstacle in the free path of a beam of light, then phenomena of interference and diffraction will occur; and Young, followed by Fresnel, showed that under these conditions Light does in fact present phenomena of interference or diffraction; while Fresnel proved, still further, that the Wave Theory of Light affords an adequate explanation of all the observed phenomena in all their details. From that moment, and throughout the rest of last century, the pure Wave Theory of Light was accepted without demur.

There exist, of course, various kinds of light, each corresponding to some definite "colour." The white light radiated, for example, by an incandescent body like the filament of an electric lamp, is formed by the superposition of a continuous sequence of simple forms of light whose colours pass by imperceptible gradations from violet to red, thus forming the spectrum. Hence the Wave Theory of Light is naturally led to associate with each kind— with each component of the spectrum—one given wave-length; in other words, one given wave-length corresponds with each colour. Since the interference phenomena depend on the wavelength, they enable us to measure the wave-lengths corresponding to the various colours of the spectrum; and it has proved possible in this way to ascertain that the wave-length varies progressively and continuously from the violet end of the spectrum, where it has the value of 4/10,oooths of a millimetre, to the red end, where it reaches 8/10,oooths.

*   *   *

We have seen, then, that some thirty years ago, no doubts were entertained but that Light, and other kinds of rays, were pure wave phenomena. Since then, however, phenomena due to radiation hitherto unknown have been discovered; and these phenomena apparently can be explained only by a corpuscular theory. The most important of these is the photo-electric effect: when (that is to say) a piece of matter, of metal for example, is illuminated, it is often observed to expel electrons in rapid motion; and observation of this phenomenon has shown that the velocity of the expelled electrons depends solely on the wave-length of the rays falling on the substance, and on the properties of this. But it depends in no way on the intensity of these incident rays: what does solely depend on this intensity is the number of electrons expelled. Further, the energy of the electrons expelled varies inversely with the length of the wave which falls on the substance in question. Consideration of this phenomenon led Einstein to grasp the fact that its explanation demanded a return, at least to some extent, to the theory of the corpuscular structure of radiation. He assumed therefore that rays are composed of corpuscles, the energy of which varies inversely with the wave-length, and has shown that the laws of the photo-electric effect follow easily once this hypothesis is adopted.

At this stage, however, physicists were in a state of no small difficulty. For, on the one hand, they had the group of diffraction and interference phenomena, which show that Light consists of waves; while on the other hand, there were the photo-electric effect and other more recently discovered phenomena, showing that Light consists of corpuscles—of photons, as they are now called.

The only way of escaping from this difficulty, then, is to assume that the wave aspect of Light, and its corpuscular aspect, are as it were two different aspects of the same underlying reality. Thus whenever a ray exchanges energy with Matter, the exchange can be described on the assumption that a photon is absorbed (or emitted) by Matter; on the other hand, if we wish to describe the motion *en masse* of light-corpuscles in space, then we must fall back on the assumption that propagation of waves is taking place. An elaboration of this idea leads to the further assumption that the density of the cloud of corpuscles, which is associated with a light-wave, is at any given point proportional to the intensity of this wave. In this way, therefore, a sort of synthesis of the two ancient rival theories is reached, so that we are enabled to explain interference phenomena as well as the photo-electric effect; but the capital interest of this synthesis consists of the fact that it shows us that, in the world of Nature, waves and corpuscles are closely interconnected—at any rate in the case of Light. And if this interconnection exists, may one not assume that it exists also for Matter? For the entire work of physicists had thus far tended to reduce Matter to a stage where it was no more than a vast collection of corpuscles. But if a photon cannot be separated from the wave which is bound up with it, then surely in the same way we are bound to assume that corpuscles of Matter are in their turn, too, universally associated with a wave. And this, in fact, is the chief question with which today we have to deal.

Let us assume, then, that corpuscles of Matter—electrons, for example—are universally accompanied by a wave. Between the corpuscle and the wave there is an intimate lien; hence the motion of the corpuscle, and that of the wave, are not independent of each other, so that a connection can now be established between the mechanical properties of the corpuscle—its momentum and its energy—on the one hand, and the characteristic values of the wave with which it is associated—its length and the velocity with which it travels—on the other. Thus on the assumption of the interconnection between the photon and its associated wave this parallelism can in fact be established: and this theory of the interconnection between the corpuscles of Matter and their associated waves is known today under the name of Wave Mechanics.

When the wave associated in this way with a corpuscle is moving freely in a region whose dimensions are great as compared with the length of the wave, the New Mechanics assigns to the corpuscle associated with the wave the motion determined by the laws of classical Mechanics. This applies particularly to the motion of electrons which we can observe directly; and this explains why observations of the large scale motions of electrons had led to their being regarded as simple corpuscles. But there are certain cases where the laws of classical Mechanics fail to describe the motion of corpuscles. The first case is one where the propagation of the associated wave is confined to a region in space having dimensions of the same order as the wave-length; and this is the case of the electrons within the atom. Here the wave associated with an electron is forced to take the form of a stationary wave, similar to the stationary elastic waves found in a cord fixed at each end, or to the stationary electric waves which may be set up in the antenna of a wireless installation. Now theory shows that these stationary waves must have certain quite definite lengths, and that in the associated electron certain equally definite energies correspond to these wave-lengths; still further, these definite states of energy in turn correspond to the states of "quantized" motion introduced into his theory by Bohr. This also furnishes an explanation for a fact which had hitherto remained extremely mysterious—the fact, namely, that quantized motion is the only type of which the electron contained within the atom is capable.

There is still another case where the electron cannot move in accord with the classical laws of Mechanics—namely where the associated wave meets with obstacles in the course of its advance. In such a case interference takes place, and the motion of the corpuscles, in relation to the motion which classical Mechanics would predict, is somewhat modified; so that to form an idea of what must then occur we may follow the analogy with rays. Let us assume, therefore, that we direct a ray of known wave-length on to an apparatus designed to give rise to interference. Since we know that the rays consist of photons, we can say that we are launching a swarm of photons upon the apparatus; and in the region where interference occurs, the photons are distributed in such a way that they are concentrated at those points where the intensity of the associated wave is greatest. Let us now suppose, still further, that we direct on the same apparatus, not a ray, but a beam of electrons having an associated wave of the same wave-length as in the previous ray. In such a case the wave will interfere as before, since it is the wave-length which controls interference phenomena. It would then be natural to assume that the electrons will be concentrated at the points of greatest intensity of the wave: in other words, that in this second experiment the electrons will be spatially distributed in the same way as that in which the photons were distributed in the first. If then it can be shown that such is in fact the case, the existence of the wave associated with the electrons will also have been demonstrated, and it will thus be possible to check the precision of the formulae of Wave Mechanics.

Now according to Wave Mechanics, a wave is associated with electrons moving with velocities usually realized experimentally, the length of the associated wave being of the same order as that of X-rays, viz. 1/10,000,000th of a millimetre. In order, therefore, to demonstrate electron-waves, we must try to produce by their means interference phenomena analogous to those obtained with X-rays; and phenomena of this type were in fact obtained—first, in 1927, by Davisson and Germer in the United States, and, later, by a great number of experimenters, among whom may be mentioned G. P. Thomson in England and Ponte in France. I shall not describe their experiments, but confine myself to saying that they ended with the complete verification of the formulae of Wave Mechanics.

These brilliant experiments have thus proved that the electron is not merely a simple corpuscle; in one sense it is at once a corpuscle and a wave. The same conclusion—as has been proved by still more recent experiments—applies to the proton. Thus we see that Matter, as well as Light, consists of both waves and corpuscles; a far greater structural resemblance than had formerly been suspected is shown to exist between Light and Matter: and our conception of Nature has thus become the simpler, and also the loftier.

* * *

The nucleus of an atom having the atomic number N has, as we saw above, a positive charge equal to N times that of the proton, and in it practically the entire mass of the atom is concentrated. It had long been believed that the nuclei of atoms consist of protons and electrons, the number of protons exceeding that of the electrons by N, and practically the entire mass being due to the protons. This idea that the nucleus is of a complex nature was more or less enforced by the interpretation of radioactivity, the discovery of which was initiated by Henri Becquerel, and in essence was the work of Pierre Curie and of his wife and collaborator Marie Sklodovska, whose death was such a grievous blow to French Science. The radioactive substances are heavy elements, bearing the highest atomic numbers in the series of elements—from 83 to 92. They are characterized by a spontaneous instability, that is by the fact that from time to time the nucleus of one of their atoms explodes, at the same time changing into the nucleus of a lighter atom. This transformation is accompanied by the expulsion of electrons (β-rays), of the light atoms of Helium ($N = 2$) (α-rays) and by extremely penetrative rays of very high frequency (γ-rays). For physicists the discovery of these radioactive phenomena was of extreme interest, since it proved to them that the nuclei are in fact complex structures, and that a complex nucleus in the process of disintegration gives rise to a simpler one—thus spontaneously realizing the transmutation of elements dreamed of by the alchemists of the Middle Ages. Unfortunately, however, radioactivity is a phenomenon on which we are unable to exert any influence, and which consequently we can merely observe without being able to modify the process. Some twenty years after the discovery of radioactivity a great step forward was taken, when Rutherford discovered artificial disintegration; for by bombarding light atoms with α-particles—which in turn are emitted by radioactive substances—it was proved possible to break up these light atoms; and in this way simpler atoms are obtained—a genuine artificial transmutation. The quantities of Matter which undergo this transmutation are naturally slight, yet it has at present substantial practical importance; theoretically, on the other hand, its interest is enormous, since it proves the unity of Matter and affords further knowledge on the structure of the nuclei.

This research into artificial transmutations has undergone considerable development in recent years, beginning in England, where, under the leadership of Rutherford, the physicists Chadwick, Cockcroft, Walton and Blackett have reached remarkable results, and later in the United States, where Lawrence's work may be mentioned. In France, Paris now possesses two very important centres where problems relating to nuclei are being pursued by young investigators of great ability. First we have the Institut du Radium, directed until her death by Madame Pierre Curie, and where Madame Joliot, née Curie, her husband Monsieur Joliot, Pierre Auger, Rosenblum and others are at work. And then there is the Laboratoire de recherches physiques sur les rayons X, founded and directed by the author's brother, where Jean Thibaud, J. J. Trillat, Leprince-Ringuet and others are, or were, pursuing skilled and fruitful studies.

I cannot here deal in any way with the details of the results obtained; these have led to a kind of nuclear chemistry, in which the transmutations are represented by means of equations strictly analogous to those long used by chemists to represent ordinary chemical reactions. I must, however, stress two fundamental discoveries made wholly unexpectedly in the course of these researches. The first of these is the discovery of the neutron; Chadwick and the Joliots independently discovered the presence, among the products of the process of disintegration, of a kind of corpuscle hitherto unknown. These corpuscles pass through Matter with great ease; they appear to have no electric charge, but to have a mass approximately equal to that of the proton. They are the neutrons, and there appears to be no doubt that they play an important part in the structure of the nuclei.

Within a year of the discovery of the neutron, in 1932, a fourth class of corpuscle was discovered in its turn. While studying the effects of the disintegration caused by cosmic rays, Anderson, and also Blackett and Occhialini, independently demonstrated the existence of positive electrons—i.e. corpuscles having the same mass as the electron and with an electric charge equal to that of the electron, but bearing an opposite sign. These positive electrons, which are a great deal rarer than the negative, appear to play an important part in the phenomena connected with the nuclei.

The upshot of these recent sensational discoveries was to leave the position a good deal more complicated than it had ever been, since we now know four different kinds of corpuscles—electrons, protons, positive electrons and neutrons. The question one asks is whether they all are in fact elementary; and the answer is undoubtedly in the negative. It would appear that one of the four must be complex. It may be assumed, for example, that the proton, the electron and the positive electron are the elementary units, in which case the neutron consists of a proton to which is due almost the entire mass of the neutron, and of an electron which neutralizes the charge of the proton. Or again one may assume—and this appears to me the more attractive hypothesis—that it is the neutron and the two kinds of electron which are the elementary corpuscles, in which case the proton would consist of a neutron and a positive electron, and would cease to rank as a simple corpuscle. In any case the discovery of the neutron and of the positive electron are valuable additions to our knowledge of the atomic world.

A word may here be said about cosmic rays. A series of experiments undertaken during recent years, the most important of which are those carried out by Millikan, has proved the existence of extremely penetrative rays which appear to come from interplanetary space. It has been found, too, that these rays have extremely powerful effects on Matter and cause various kinds of atomic disintegration. Research into cosmic rays is difficult, and as yet little is known of their nature; but there is small doubt that numerous interesting results will shortly be obtained in this respect as well.

*   *   *

All too brief as this survey is, it will have shown that laboratory research during the last few years has led to results of the utmost interest almost each day. But theoretical Physics, too, whose function it is to provide a guiding light for experimental Physics, has not remained idle.

In the history of theoretical Physics, then, during the last thirty years, there are two great landmarks: the Theory of Relativity, and the Quantum Theory, two doctrines of the widest scope; and while the Theory of Relativity is less closely connected with the advancement of atomic Physics, it is the more familiar to the man in the street. Its origin lies in certain phenomena of the propagation of Light which could not be explained by the older theories; but by an intellectual effort which will always hold an eminent place in the annals of Science, Einstein removed the difficulty by the introduction of entirely novel ideas on the nature of Space and Time and their interrelation. Hence the origin of that remarkable Theory of Relativity, which later achieved an even more general scope by providing us with an entirely new conception of Gravitation. It is true that certain of the experimental verifications of the Theory have been, and still remain, in debate; but it is quite certain that it provides us with extremely novel and fertile points of view. For it has shown how the removal of certain preconceived ideas, adopted through habit rather than logic, made it possible to overcome obstacles regarded as insuperable and thus to discover unexpected horizons; and for physicists the Theory of Relativity has been a marvellous exercise in overcoming mental rigidity.

The Quantum Theory and its developments, if less generally familiar, are certainly at least equally important, since by means of this Theory it has been possible to make use of the discoveries of experimental Physics to form a science of atomic phenomena. When a more precise description of these phenomena was felt to be necessary, the fundamental fact which became apparent was that it was imperative to introduce completely novel concepts which had been entirely unknown to classical Physics. For in order to describe the atomic world it is not enough to transport the methods and images which are valid on the human, or on the astronomical scale, to another and very much smaller scale. We saw that, following Bohr, scientists succeeded in imagining atoms to be miniature solar systems in which the electrons played the part of the planets, and in tracing their orbits round a central sun bearing a positive charge. But if this image was to give really valuable results, it became necessary to assume, still further, that the atomic solar system obeyed Quantum Laws; and these were entirely different from the Laws governing the systems with which Astronomy deals. The more carefully this difference was considered, again, the more its wide scope and fundamental significance began to be appreciated; for the intervention of quanta brought about the introduction of discontinuity in atomic Physics, and this introduction is of essential importance, since without it atoms would be unstable, and Matter could not exist.

We saw that the discovery of the double nature of electrons, as at once corpuscular and undulatory, was followed by a change in the Quantum Theory, so that this was given a new form, some years ago, called Wave Mechanics. The new form has met with manifold success, and Wave Mechanics has brought about a better understanding and prediction of those phenomena which depend upon the existence of quantized stationary states for atoms. Every branch of Science, including Chemistry, has benefited from the impetus due to the new theory, because this has brought with it an entirely novel and interesting manner of interpreting chemical combinations.

The development of Wave Mechanics, then, has compelled physicists to give an ever wider and wider scope to their concepts. For according to the new principles, the Laws of Nature no longer have the strict character which they bear in classical Physics: phenomena (in other terms) are no longer subject to a rigorous Determinism; they only obey the Laws of Probability. The famous Principle of Uncertainty advanced by Heisenberg gives an exact formulation to this fact. Even the notions of Causality and of Individuality have had to undergo a fresh scrutiny, and it seems certain that this major crisis, affecting the guiding principles of our physical concepts, will be the source of philosophical consequences which cannot yet be clearly perceived.

MATTER AND LIGHT IN MODERN PHYSICS

"Know then that bodies have, as we call them, their semblances which are slender membranes detaching themselves from their surface and flying in every direction in the air. . . . These semblances must traverse incalculable distances in a flash; first, because they are exceedingly small elements, and there is behind them a cause which thrusts them forward; and secondly they fly in swarms so subtle that they can easily penetrate and as it were pour through the air."

Such are the words in which Lucretius, in his *De Rerum Natura* (Book IV), summed up his theories of Light 2,000 years ago. For the thinkers of antiquity it was not easy to understand how an image can be formed in the eye; and it must be admitted that it was an ingenious solution to imagine that every element at the surface of a body emitted a small particle which was the faithful reproduction of that element. The theory was that the particles, having passed over a greater or less distance, would unite within the eye, there forming the exact image of the body from which they had been emitted, just as the stones of a mosaic would reproduce the original pattern if they were reunited after momentary scattering. Lucretius had seen that such a theory would imply that an "incalculable" velocity would have to be attributed to the particles of Light; and there is cause for amazement at the profundity of this intuition, if we reflect that the first exact experiments, which attributed to Light the enormous speed of 300,000 kilometres a second, were not made until the middle of the nineteenth century.

Fundamentally, however, Lucretius' theory of Light was one aspect of a general doctrine—of the Atomic Theory—which was the delight of the ancient philosophers. According to this Theory,

all natural phenomena can be explained by the motion and interactions of indivisible corpuscles, that is of atoms in the etymological
sense of the word. These were supposed to be the ultimate reality
beyond which scientific analysis was not required to penetrate,
since after all a halt must be made somewhere—ἀνάγκη στῆναι.
All substances therefore, whether solid, liquid or gaseous, must be
made up of such atoms, and even Light, less different in this
respect from Matter than a superficial analysis might lead the
observer to think, must be formed of corpuscles which, while
certainly lighter, swifter and subtler than the others, were not in
their essential nature different from them.

*   *   *

A long series of centuries must now be passed over at a step;
we are now to consider what was the general state of our knowledge of physical phenomena some thirty years ago. First we must
note the profoundly important differences which separate modern
from ancient Physics. The latter was based on certain correct
observations and also on a number of naive legends, on which
were built theories which were too ambitious and at the same time
wholly qualitative. Very different is the character of modern
Physics, which is based on experiment. This is a more exact method
than that of simple observation; it is a method which deliberately
produces certain given conditions in order to see what phenomena
are caused by these conditions. Then, having thus established
beyond dispute relations of interdependence among physical
facts, it tries to interpret these relations and to find a place for them
in a single general schema, its means to this end being the use of
an extremely powerful intellectual instrument, Mathematical
Analysis, of which the ancients did not know the use. In other
words, the experimental method allows certain Laws to be established. Theory then interprets these Laws by establishing a connection with a single principle; and finally it uses the principle
thus formulated in order to predict quantitatively other phenomena. Finally, recourse is had once again to experiment to verify
the precision of the predictions made by Theory.

Thus the exactness of their methods, and the certainty of their conclusions, had rendered the physical sciences about the year 1900 infinitely superior to the vague outline with which Lucretius and his contemporaries had been forced to remain satisfied. On the other hand, Lucretius' fundamental idea that the phenomena of Nature were explicable by the motion and interactions of elementary particles, so far from having been abandoned, had been brilliantly confirmed by experiment. Following Lavoisier, chemists had established the fact that there exist a certain number of simple substances, or elements, which resisted every attempt to decompose them, and that by suitable means every other chemical substance can be decomposed into a certain number of elements. In the next place, as has already been remarked in the preceding Chapter, it had been found possible to account for all the general Laws of Chemistry by assuming that simple substances consist of atoms, each substance having its own special kind of atom. On this view all other chemical substances consist of molecules, which are more or less complicated combinations of atoms of which simple substances are built up, while chemical analysis consists in disintegrating these molecules, thus setting free the elements from the combination in which they were held together.

The Atomic Theory, triumphant in Chemistry, had also guided physicists in their views of Matter and its properties. At the end of the seventeenth century Newton—aided by the recent progress of the mathematical sciences to which he himself had so powerfully contributed—had laid the foundation of Mechanics, as the science which enunciates the laws in accordance with which a particle or group of particles moves under the influence of given forces. By adding to these laws the hypothesis that all particles of Matter are attracted to each other with a force varying directly as their masses, and inversely as the square of their distances (The Universal Law of Gravitation), Newton formulated the science of the motion of the heavenly bodies, or Celestial Mechanics; a science so exact that, as we know, Leverrier was able to discover by calculation alone the existence and the position of a planet (Neptune) hitherto unknown.

Armed in this way with the laws of Mechanics, physicists were in a position to attempt the explanation of all the properties of Matter on the hypothesis that they are due to the motion and the interaction of elementary particles—that is of atoms; and during the course of the nineteenth century these attempts met with a considerable degree of success.

Thus to explain, for example, why the surface of a liquid contained in a narrow tube, instead of being flat, shows in certain cases a curve giving it the aspect of a convex, or in other instances of a concave meniscus, the capillary theory, due to Laplace, attributes these phenomena to the forces at work between the atoms of the solid matter forming the tube, and the atoms of the liquid, and succeeds in predicting exactly all the laws governing these capillary phenomena.

Again, to predict the properties of gases, Maxwell and Boltzmann's kinetic theory tells us that the gases are formed of atoms in rapid motion, that the pressure of a gas on the walls of the containing vessel is due to the impact of the atoms against the wall, and that the temperature of the gas is the measure of the mean energy of the atoms' motion; and this admirable theory explains quantitatively a whole series of properties of gases—both such as had already been established by experiment and such as were proved later.

Once again, with regard to the thermal properties of solids, the Atomic Theory considers these as formed of atoms in equilibrium, but capable of oscillating about this position with an intensity increasing with the temperature of the body. It also calculates the amount of heat required by a given mass (say one gramme) in order to raise its temperature one degree Centigrade: and the law which it thus discovers agrees exactly with that which the French physicists Dulong and Petit had derived from their experiments.

Many more instances might be adduced; for the atomic view has provided very satisfactory interpretations of the properties of material substances. For although the atoms are so small and so extremely light, their disorderly motion in gases has been caused to manifest itself directly, and their masses and number in a given volume have been measured. Thus we know today the number of atoms in a cubic centimetre of gas, even though it may exceed a milliard milliards!

*   *   *

A fresh success was won by the Atomic Theory when Sir J. J. Thomson, and later H. A. Lorentz, succeeded in explaining the entirety of electromagnetic phenomena by the hypothesis that electricity has an atomic structure. Electric and magnetic phenomena only began to be studied systematically towards the end of the eighteenth century, but made very rapid progress in the first half of the nineteenth century, thanks particularly to the memorable work done by Volta, Coulomb, Ampère, Biot, Laplace and others. These phenomena can be interpreted by the two-fluid theory referred to in Chap. I. If a substance contains equal quantities of these two fluids it is electrically neutral: if it contains more positive than negative fluid, it has a positive charge; and vice versa. All electrostatic phenomena can be explained in this way, while electric currents are interpreted as the motion of a certain quantity of the two fluids. The fundamental idea introduced by J. J. Thomson and H. A. Lorentz is that the negative electric fluid (as I have already observed)[1] is always formed of particles which are all alike, having an extremely small mass and bearing the same negative and extremely minute charge—the electrons. Later, scientists have been led to generalize this hypothesis by assuming that the positive fluid consists *inter alia* of particles previously described as protons.

In the skilful hands of Lorentz and his school, still further, the theory of electrons has succeeded in explaining electromagnetic phenomena which were already known, and also in predicting new ones. But the actual existence of electrons has also been demonstrated more directly. Thus their mass and charge—inconceivably small quantities—have been measured; the fact has been verified that the laws of Mechanics apply to them, and they have even been collected, one by one as they arrive, in an apparatus which

[1] P. 21.

emits a sound as each enters, like a resounding target producing a note at the successive arrival of the projectiles constituting a salvo.

The discovery of electrons and protons gave to physicists the agreeable, but nevertheless possibly deceptive, impression that they had found the ultimate elements of Matter. Actually, before their discovery it was the atoms of simple substances which had to be considered as the elements of which Matter was built up. Unfortunately there are a great number of simple substances, and the human mind, in its desire for simplicity, refuses to remain satisfied with accepting the existence of so many ultimate elements. Modern theories, however, on the details of which I cannot here insist, have led us to regard the atoms of simple substances as being complex systems consisting of protons and electrons, and hence to reduce the ultimate elements of Matter to two.[1] Let us (for example) consider the lightest element, Hydrogen. According to Bohr the Hydrogen atom consists of a proton around which an electron revolves, the whole system being electrically neutral, since the proton and the electron each possess equal charges of opposite sign; and there are indeed a large number of facts to substantiate the idea that all atoms are structures composed of protons and electrons.

Thus it might have seemed that the ideal of the ancient philosophers had almost been attained, since the hypothesis of the existence of only two sorts of corpuscles allowed all the properties of Matter to be inferred from the motion and interactions of these corpuscles. In the preceding exposition, however, I have confined myself to only one of the two branches of modern Physics—that which deals with the properties of Matter. But we must now consider the manner in which the other main branch of Physics has evolved—that dealing with the properties of Light, or, as it is put today in a more general way, with the properties of radiation.

*   *   *

The nineteenth century had witnessed the triumph of the Atomic Theory as applied to the Physics of Matter; but it had

[1] But cf. further pp. 75 ff.

witnessed its failure as an explanation of the phenomena of Light. The idea that Light consists of corpuscles in a state of rapid motion had found an illustrious champion in Newton, who had shown that this idea permitted an interpretation of the leading optical facts known to his own period: the propagation of Light in a straight line, its reflection in mirrors and its refraction in transparent media. The entire eighteenth century, bowing before Newton's authority, accepted his view, which even at the beginning of the nineteenth century was finding eager defenders. At this point, however, the brilliant work of the French scientist Fresnel caused the complete abandonment of the views held by Lucretius and Newton.

I have already observed that as early as the seventeenth century, in fact, a contemporary of Newton's earlier period—Huyghens—had put forward a theory of Light which differed completely from the corpuscular theory.[1] Accordingly Light, instead of having a discontinuous structure and consisting of elements which preserve their individuality, would be a disturbance propagated through space. Huyghens developed his notion with great ingenuity and drew conclusions from it which, having been rehabilitated by Fresnel, are still enunciated in our treatises on optics. It is true that Newton, with characteristic profundity of thought, had been tempted at times to combine his own corpuscular theory with the undulatory view; but on the whole his authority, as we saw, contributed to the abandonment of Huyghens' theory.

Attention was concentrated once again on the Wave Theory in consequence of the discovery of the phenomena of interference and diffraction, which had originally been initiated by Newton himself, but owed most to the work of Young, confirmed and extended later by the work of Fresnel. The phenomena due to a wave moving without meeting an obstacle have previously been discussed,[2] together with what happens when a wave reaches a region in space where there are scattered obstacles. If then Light consists of waves, it should be possible to obtain perfectly characteristic phenomena when it falls on mirrors, screens or other

[1] P. 25.   [2] P. 26.

similar obstacles. There will be places where the waves will annihilate each other, so that there will be dark zones—dark fringes, as physicists call them; elsewhere on the other hand the waves will intensify each other, and the luminosity will be concentrated, so that there will be bright fringes. These are the delicate phenomena, often difficult to observe, whose existence Young and Fresnel were able to establish beyond doubt, the latter further achieving the honour of demonstrating that the Wave Theory furnishes a complete explanation for them, whereas corpuscles moving in accordance with the Newtonian laws of Mechanics could never produce such phenomena. After Fresnel, therefore, the question appeared to have been solved: Light consists of waves.

But during the last hundred years physicists have succeeded in discovering a whole series of "invisible" rays which do not affect the human retina. Among these are, in the first place, the rays belonging to the infra-red end of the spectrum having a wave-length greater than that of red; and the rays belonging to the ultra-violet end, whose wave-lengths are shorter than those of violet. Still further, there are the electromagnetic waves (which are used in wireless transmission) which are a good deal longer than the infra-red, and, at the other end of the scale, there are the X-rays and the γ-rays emitted by radioactive bodies whose waves are incomparably shorter than those of the ultra-violet rays.

For some time it seemed as though the properties of all these types of radiation could be explained solely on the wave hypothesis. Physics was thus divided into two separate branches: the Physics of Matter, where the hypothesis that the particles, of which the elements are constituted, move in accordance with the laws of Newtonian Mechanics allowed all the known facts to be co-ordinated; and the Physics of radiation, where all the facts were interpreted by wave motion. We shall see, however, that recent discoveries have eliminated this partition, by which Physics was thus divided, and have shown the necessity of a synthesis whose intellectual appeal would give unmixed pleasure to the scientists

of our day, were it not that its final perfection is excessively difficult
of attainment.

* * *

The new facts which have been discovered within the last 25
years, and the new knowledge which has compelled us to recast
our views, are concerned with Quantum Phenomena; and the
honour of having been the first to suspect their existence is due to
Planck. Without entering into any detailed description, I should
like at this point to give a general idea about them by showing
how some of them have revealed the corpuscular structure of
Light, while others have shown how frequently traditional
Mechanics was unable to predict the motion of the ultimate
corpuscles of which Matter is made up.

The leading phenomenon to draw attention to the possibility
that Light has a discontinuous structure is the photo-electric effect.
Let us consider a source of light. According to Fresnel's Wave
Theory this source emits a spherical wave which spreads through
the ether. Hence the energy emitted by the source is disseminated
in space, and the effect which Light can produce will be the weaker
in proportion to the distance of a given point from the source.
According to the corpuscular theory, on the other hand, the
source sends out particles in every direction, and these particles are
not scattered. They remain indivisible units, and can hence produce
a considerable effect even at a great distance. Now the discovery
of the photo-electric effect has indubitably shown that every type
of radiation produces considerable effects on Matter, and that these
effects *do not diminish as the distance from the source increases*. It
was, then, by such facts that Einstein found himself compelled to
return to some extent to the corpuscular theory, by assuming that
the radiant energy is divided into small particles or corpuscles,
and that these are emitted in all directions by any given light-
source. Thus a third kind of corpuscle was introduced, in addition
to electrons and protons—light-quanta or, as it is more usual to
say today, photons.

Another phenomenon, which was discovered in 1922 by the

American physicist H. A. Compton, provides further proof in the same direction. The action of radiation upon an isolated electron is the same as though there were a collision between two corpuscles; a fact which once again leads towards the concept of corpuscles of radiation. On the other hand, since the phenomena of interference and diffraction, which were the starting-point whence Fresnel inferred the necessity of the Wave Theory, are also in need of interpretation, it follows that the quantum phenomena clearly showed the necessity of discovering synthetic principles within the sphere of radiation, or a theory finding room for the point of view of Fresnel and for that of Newton simultaneously.

Now we have seen that modern physicists tried to envisage Matter as consisting of two kinds of elementary particles—electrons and protons. But when it was sought to recognize in these the qualities of material atoms by supposing that they consisted of a central nucleus, and of a group of electrons revolving around this nucleus like the planets around the sun, it was found necessary to modify the classical laws of Mechanics in a quite unexpected manner. Thus the outlines of a new Mechanics—Quantum Mechanics—were next formulated, thanks particularly to the efforts of Planck, Bohr and Sommerfeld; but at the same time, the reason why this modification of Mechanics had to be introduced for the particles within the atom remained for long extremely obscure.

* * *

A consideration of these problems led me, in 1923, to the conviction that in the theory of Matter, as in the theory of radiation, it was essential to consider corpuscles and waves simultaneously if it were desired to reach a single theory, permitting of the *simultaneous* interpretation of the properties of Light and of those of Matter. It then becomes clear at once that, in order to predict the motion of a corpuscle, it is necessary to construct a new Mechanics—Wave Mechanics as it is called today—a theory closely related to that dealing with wave phenomena, and one in which the motion of a corpuscle is inferred from the motion in space of a wave. In this way there will be, for example, Light corpuscles, photons; but their motion will be connected with that of Fresnel's waves, and will provide an explanation of the phenomena of interference and diffraction. Meanwhile it will no longer be possible to consider the material corpuscles, electrons and protons, in isolation: it will, on the contrary, have to be assumed in each case that they are accompanied by a wave which is bound up with their own motion. I have even been able to state in advance the wave-length of the associated wave belonging to an electron having a given velocity.

Newtonian Mechanics undeniably succeeds in predicting exactly motion occurring on the human scale, and also on the scale of celestial bodies; and the reason for this is that, for Wave Mechanics, Newtonian Mechanics is an entirely adequate approximation. But when it comes to investigating the motion of the material particle inside the atom, the old Mechanics ceases to have any value, while the new one allows us to grasp the sense of the new principles which the Quantum Theories were obliged to introduce. In an admirable series of Papers, Schrödinger has given exact form to the application of these ideas, and has shown how they help to furnish a complete justification of the generalized form given a little earlier to Quantum Mechanics by the young German physicist Heisenberg.

The synthetic theory has thus had complete success; what was still lacking, however, was a set of direct experiments proving the existence of the associated wave—the wave (that is to say) associated with the material particles. I have already observed that two American physicists, Davisson and Germer, and two British physicists, G. P. Thomson and Reid, succeeded in obtaining interference phenomena entirely analogous to those obtained in the case of radiation, by utilizing beams of electrons. A study of the phenomena observed enables the length of the wave associated with the electrons employed in the experiment to be calculated, and it has been shown that this length agrees exactly with the figure I had predicted three years earlier. Wave Mechanics has thus played the part essential to all sound physical theory: it has predicted phenomena the existence of which has later been demonstrated experimentally.

But, at the same time, there was no occasion to believe that all difficulties had vanished, and that physicists would now be able to advance along a single and smooth road. For it seems to be definitely established that both Matter and radiation have two aspects—that they may be regarded either as wave, or as corpuscle; and still further, that if this duality is allowed for, all Physics might be united into a single whole through the instrumentality of one synthetizing theory. Nevertheless, the reason why these two aspects exist, and the manner in which it might be possible to merge them in one superior unity, remain a matter of mystery. On these points the opinions of the best qualified scientists still continue to differ widely, and even the principles on which the scientific explanations have hitherto been based have been subjected to severe criticism, the outcome of which it is too early to foretell.

Such difficulties cannot, however, be matter for surprise. For whenever the human mind has laboriously succeeded in deciphering one page in the Book of Nature, it has seen immediately how much greater must be the difficulty of deciphering the following page. Yet a deep instinct forbids it to feel discouraged, and urges it towards a renewed effort to penetrate further and further in the knowledge of the harmonies of Nature.

3

QUANTA AND WAVE MECHANICS

It is strange to observe how quanta, having first made their way into the Theory of radiation, have later invaded the whole of Physics. In the beginning—about 1900—they played a very modest part: in the Theory of the exchange of energy between Matter and radiation it was seen that the energy of the oscillating material entities, on the atomic scale, was always equal to an integral multiple of a certain quantum of energy given by the product of the frequency of the oscillator into Planck's constant, h. But it was quickly found that if this was true of the elementary material oscillators, then the same principle must apply to all motion of very small scale corpuscles, and also that the latter must submit to similar restrictions. In fact, the only periodic corpuscular motion which is stable, and actually realized in Nature, is that fulfilling the condition that the "mechanical action" calculated for any complete period is equal to a multiple of Planck's constant, h; and hence this constant is found to play the part of a quantum of action, the quantum of energy of the oscillator being merely the outcome, in any given instance, of the existence of this quantum of action.

Thus quanta found their place in Mechanics, where they restrict the motion that is possible by rules of quantization, in which integers find a place; and the appearance of these integers in the dynamics of elementary corpuscles seems, at first sight, a somewhat strange fact.

But with the Theory of light-quanta, the following years brought a surprise of a very different kind. For the discovery of the photoelectric effect had shown that the absorption by Matter of radiant energy, of frequency v, occurs in finite quantities equal to hv. Reverting to the corpuscular theories of Light favoured by the

eighteenth-century scientists, therefore, Einstein now assumed
that Light of frequency ν was formed of atoms of energy having
the value $h\nu$. This hypothesis then allowed him to formulate a
general law of the photo-electric effect, with which the facts were
found exactly to agree. In this way the Wave Theory of Light,
which since the days of Fresnel had held the field, was suddenly
shaken: for this theory implies that energy is distributed continuously over a luminous wave, and thus denies the possibility
of light-quanta. And yet the Wave Theory was founded on
indubitable demonstration, since the phenomena of interference
and of diffraction are interpreted by it, and by it alone. Yet it
seemed, on the other hand, that the photo-electric effect could
only be interpreted by some sort of corpuscular concept. The
contradiction was disturbing, and it did not seem easy to
eliminate it.

The years following on the introduction of light-quanta did
nothing to resolve the mystery. The importance of quanta has
continually been demonstrated. The quantum of action has been
met with in a thousand phenomena, and the constant, h, has been
measured in more ways than one. Then came Bohr's theory to
show that if the atoms of Matter are in a stable state, the reason is
to be looked for in the quantized character of the intra-atomic
motion of the corpuscles. Since then, indeed, we have learned that
the stability of Matter and its very existence rest upon quanta.
Certainly this fact gave us the measure of our ignorance before we
knew about quanta: but it did not help us to a better understanding
of their true nature. And while the importance of quanta for small-
scale motion thus became more and more indubitable, the corpuscular structure of radiation—i.e. the existence of photons, as we
express it today—was confirmed by research into the photo-
electric effect for X-rays, and also by the discovery of the Compton
effect.

Such was the position about 1924. Since then, however, it has
undergone a considerable evolution. It would be incorrect, of
course, to say that we now understand completely the innermost
meaning of quanta: nevertheless we have made advances in that

direction. The starting-point of this progress is to be found in an idea which was the basis for the new Wave Mechanics. It is as follows. To explain the elementary properties of Matter and of radiation, the idea of waves and that of corpuscles must be made use of *simultaneously*; and the magnitudes which express these two ideas mathematically are interconnected by relations in which Planck's constant invariably has a place. Each time that the existence of Matter, or of radiation, is manifested to us by some elementary phenomenon, whatever happens is compatible with the existence of elementary corpuscles both of Matter and of Light, which is the reason why experimental research has shown corpuscles to exist on each occasion when it has been able to investigate any elementary phenomenon. For Matter, such corpuscles were electrons and protons, and for Light, photons. But the new Mechanics has shown that it is impossible to follow continuously the individual activities of these corpuscles: the only predictions possible about this activity are of a statistical nature and, in order to obtain them at all, it is essential to make use of the idea of waves. In other words, it is by means of waves that we can predict statistically the distribution and motion of corpuscles.

The way in which what I have said applies to Light, and to radiation in general, will now no doubt be readily understood. On the one hand, investigation of the elementary phenomena in which radiation acts on Matter (like the photo-electric effect and the Compton effect) has given us a proof of the fact that when radiation of frequency v acts on Matter, things happen exactly as though the ray consisted of corpuscles having energy equal to hv. On the other hand, interference and the diffraction of Light have been known to exist for a long time, and these show us that the mean radiant energy is distributed in accordance with the Wave Theory. It is thus natural to assume that there are corpuscles of radiant energy—photons—each with energy equal to hv, whose activities, as Time goes on, are represented statistically by Fresnel's classical light-wave, though it is quite impossible to give in full detail the individual history of each photon. Such is the original view-point of Wave Mechanics with regard to Light; and it undoubtedly succeeds in reconciling the existence of the photo-electric effect with that of the phenomena of interference.

But the question now arises how this same general conception is to be adapted to Matter. Here too the corpuscles dealt with—electrons and protons—have long been familiar, while others, such as neutrons and positive electrons, are more recent discoveries. But what about waves? At first sight indeed it might seem as though waves are here quite superfluous, since in the electric or magnetic fields, which we can artificially create, the electrified corpuscles behave as though they were respectable corpuscles of the classical type, obeying the laws of traditional Mechanics of the material point. But in fact we saw how the development of the Quantum Theory showed that affairs are not so simple on the atomic scale, where the constant h intervenes and restricts the possibilities of corpuscular motion. Integers also intervene at this point. Now the intervention of integers is something which the classical dynamics of the material point cannot in the least explain, though it is of common occurrence in wave theories—for example in the formulae relating to interference or resonance phenomena. This naturally leads us to believe that in dealing with Matter too we must take into consideration waves, whose motion regulates that of corpuscles, if only statistically. In the Theory of Matter, in fact, this introduction of waves side by side with corpuscles has proved quite successful, since among other things it has supplied us with an explanation of the quantized character of corpuscular motion on the atomic scale. What it has shown us is, in fact, that in order to be stable the motion of a small-scale corpuscle must be associated with a stationary wave. As a rule this condition is not fulfilled, which explains why only some, but not all, motion is stable.

The new Theory further enables us to explain why the electrons in a large-scale electric or magnetic field behave like corpuscles of the classic type; and it also enables us to predict phenomena having an altogether novel character. For the light-wave governs the spatial distribution of the photons associated with it; further, when interference and diffraction occur, the photons are spatially localized in proportion to the intensity of the light-wave at any given point: and consequently one would expect that something analogous would apply to material corpuscles. One would expect that if the propagation of the wave, associated with a flow of material corpuscles having the same energy, gives rise to interference, then the corpuscles in question ought to be spatially distributed in proportion to the intensity of the wave, and should give rise to phenomena which the old dynamics of material points was quite incapable of predicting. In fact, this bold theoretical prediction has been verified by experiment. For according to Wave Mechanics, if a parallel beam of electrons having the same energy is directed on a crystal, the wave which directs the motion of the electrons will be scattered by the regularly distributed centres of the crystal lattice. The shorter scattered waves will interfere with each other, and this interference will produce maxima of scattering in certain directions; and these directions can easily be calculated if the constants of the crystal lattice employed and the length of the incident wave are known. In the new Theory, then, everything happens as though the wave controlled the entire motion of the corpuscles, and hence the electrons scattered by the crystal are bound to concentrate in those privileged directions of scattering just mentioned. And this is precisely what experiments have shown, thus affording a striking confirmation of the new ideas. This confirmation, still further, has been quantitative, since the exactitude of the fundamental equation of Wave Mechanics—$\lambda = \dfrac{h}{mv}$—was verified with great precision. This equation gives the length of the wave, λ, associated with a corpuscle of mass m and velocity v, by introducing the Quantum constant h.

In this way, therefore, the new Wave and Quantum Mechanics was established on a solid experimental basis. It has taught us to consider Planck's constant, h, as a kind of connecting link between the idea of waves and that of corpuscles. And because the constant h has a finite value, the two ideas are both necessary, while their respective validities restrict each other.

There is one other interesting point which I should like to

emphasize. If the constant h were infinitely small, the light-quanta having the value $h\nu$ would also be infinitely small, and their number in a ray of given energy would be infinitely great. Everything would then occur as though the rays had a continuous structure, and were of that purely undulatory character which Fresnel and his school attributed to them. On the other hand, the material corpuscles could in such a case easily be proved to obey exactly the classical laws of the dynamics of the material point, and there would be no need whatever to introduce waves into the Theory of Matter. Hence, in other words, if the value of h were infinitely small, classical Physics would be absolutely valid. But if, on the other hand, Planck's constant were infinitely great then light-quanta would be enormous, and their existence would leap to the eye—if I may so put it—of the least attentive physicist. In this case, however, the material corpuscles would never again follow the laws of classical dynamics, and as soon as they began to be investigated at all it would be perceived that a wave must be introduced to predict their motion. Now in Nature, such as it actually is, the constant h is neither infinitely great nor infinitely small. Actually its value is finite, and from the human point of view it appears exceedingly small, since, expressed in centimetre-gramme-second units it is given by the number $6 \cdot 55 \times 10^{-27}$. Hence for the human observer, an infinitely small h is much nearer to the actually existing conditions than an infinitely great one; and this simple truth explains the real meaning of the recent developments in Physics. For we can now understand how it is that the Physics of yesterday was induced—after a rather cursory examination—to maintain that Light had a continuous structure and the character of a wave, while at the same time attributing to Matter a discontinuous structure and asserting that it was built up of corpuscles obeying the classical laws of dynamics. It required the far more ingenious and delicate experiments made by contemporary physicists to disclose the other aspect of the facts—I mean the discontinuous aspect of Light and the undulatory aspect of Matter.

II

MATTER AND ELECTRICITY

I

A DISCUSSION OF THE CONCEPT OF THE WAVE AND THE CORPUSCLE[1]

WHEN physicists speak of energy being transferred to a distance through a certain medium, two ideas come before their minds: that of waves, and that of corpuscles; but, as happens so often, the meaning of the concepts behind these two words has undergone a considerable p. cess of evolution, so that today a careful examination of the meaning we attach to these two words is by no means s" rfluous.

M. nics attempts to predict the motion of a "material point" acted upon by forces. Naturally the notion of a material point implies no more reality than that of a geometrical point. When therefore we begin to envisage concrete instances the material point becomes a corpuscle of some kind—either an atom, an electron or one of the new components of the nuclei like those neutrons which have only recently found a place in Physics.

Optics, on the other hand, has successively adopted the theories of emission and of undulation, and has attained today—owing to quanta phenomena and the establishment of the photo-electric effect—to the notion of a kind of light-corpuscle, the photon, the nature of which still remains somewhat obscure.

Wave Mechanics, finally, associated with the motion of any given "projectile" the idea of a wave; a wave, certainly, having no physical reality whatever, but nevertheless enabling the motion of the moving entity to be predicted as exactly as possible. In this way Wave Mechanics showed that it was quite meaningless to contrast the wave aspect and the corpuscular aspect, and that with

[1] Written in collaboration with M. Maurice de Broglie.

any given phenomenon it was necessary to take both points of view into consideration to some degree.

*   *   *

Contemporary atomic Physics, then, has placed in the limelight, and has investigated, various corpuscles having properties which in each case cause a mechanical mass to be attributed to them, and in most cases also a positive or negative electric charge. The effect of an electric charge is that when the moving entity is travelling rapidly through a gas, its trajectory is traced out by a certain number of ions. Hitherto this property has not characterized the path of radiation, although the Compton effect does detect sources of ionization at certain intervals along the ray's trajectory.

On the other hand, waves do resemble neutral corpuscles in a state of motion, for in both cases the trajectories are invisible. Energy is dissipated particularly on the occasion of such exceptional encounters as produce the photo-electric effect, or again the projection of recoil nuclei.

If the mass of the proton is taken as unity, we know that the mass of the electron at rest is rather more than 1/2,000th, while that of the neutron is probably very near unity. Any moving entity with a mass either approximating to 0, or at any rate of an order of magnitude very much smaller than the preceding values, would form a category of corpuscles having very different properties. It could, for one thing, have no energy; that is, it could not reveal itself in any way unless its velocity were very near that of Light, the latter, according to current theory, forming a limit impossible to exceed; and we shall see below that these are precisely the characteristics which should be attributed to photons.

*   *   *

Among the examples which may serve to illustrate the ideas we are now developing is the following. It often happens in modern research that we have to ask if a given effect is to be attributed to a corpuscle or to some type of radiation (i.e. a photon). What is the exact meaning of this question? And further, are its terms stated in the most appropriate manner?

A reply is not always easy; and in order to get to the bottom of the difficulties it is by no means superfluous to consider rather more closely the way in which the problem of the corpuscle and its associated wave becomes formulated in the case of the photon, and more particularly to examine the relation between the wave associated with the light-corpuscle and the classical wave of optics.

If it were possible to say with absolute accuracy that the photon is a corpuscular projectile like all the others, having an extremely minute mass when at rest and hence a velocity always approximating very closely to that of Light: and if it were possible to add that the classical light-wave is identical with its associated wave, of which the velocity of propagation is known to be always greater than that of Light, and to approximate to that velocity when the velocity of the moving entity itself approximates to it—if it were possible to make these statements, then reconciliation between optics and Mechanics would have been achieved, and the only difference we should have to look for between photons and corpuscles would be that between the orders of magnitude of their respective masses.

Unfortunately it is impossible to go quite as far as that. To make it possible to regard the photon as the limiting case of a corpuscle whose mass tends towards zero, and where the associated wave is identical with the light-wave as defined by the classical fields of Maxwell and Lorentz, it would be necessary that these two mathematical entities could be reduced to each other, and should in particular possess the same elements of symmetry; but this is not the case.

The Ψ wave of primitive Wave Mechanics is a scalar magnitude, and it cannot be identified with the electromagnetic wave represented by two vectors. Dirac's theory of the magnetic electron has introduced a Ψ wave of four components; but these four components too lack the character of the components of a vector, and this forbids us to identify them with the fields already mentioned, despite the fruitless attempts which have been made in this direction.

It is possible that we may hope to find some connection between the light-fields of classical theory and the associated wave, by means of which the former could be derived from the latter, despite the difference in the elements of symmetry. This bridge, however, has yet to be built;[1] and in any case there could be no *identity* between the electro-magnetic light-wave and the associated wave having photons for corpuscles.

* * *

The concept of the corpuscle also requires a certain amount of attention.

By a corpuscle we mean a manifestation of energy, or of momentum, localized in an extremely minute volume and capable of moving through space with a finite velocity; while if the postulates of Relativity are admitted, this velocity must always be less than that of Light.

It must further be added, if we wish to adhere to the customary definition of a corpuscle—which is derived from our knowledge of classical dynamics—that it must be possible to follow the history of a given corpuscle through Time and Space and to trace it exactly and continuously; or at any rate with such approximations as the Principle of Indeterminacy allows.

The personality—to put it that way—of the corpuscle is thus preserved (although at the instant where interaction with other particles occurs, the current aspect of our theories appears under a "totalitarian" form where the distinction between individuals plays no part); and, further, it should be possible to make calculations respecting these minute constituents of Matter.

In order to interpret the anomalies of the Zeeman effect, however, and also the fine structure of the optical and the Röntgen spectra of the elements, recourse was had to the device of endowing

---

[1] Since this was written the problem has been partly solved by research pursued by the author and summed up below at the end of the chapter entitled: "Old Ways and New Perspectives in the Theory of Light."

the electron with the new property of Spin. An idea of the spin of the electron, in accordance with classical concepts, can be formed by imagining the latter to be a small sphere of electricity in rotation on one of its diameters, and by assuming that the magnetic moment set up by this rotation is equal to one Bohr magneton, while the corresponding moment of momentum is equal to half the quantum unit $\frac{h}{2\pi}$. Nevertheless this classical idea, like every idea of the kind when applied to phenomena on the atomic scale, does not seem to express exactly the real essence of spin. Spin should rather be considered as an intrinsic property of the entity known as an electron—a property to which there corresponds the existence of a magnitude having the physical character of magnetic moment, and also of another magnitude, closely linked to the first, possessing the physical character of moment of rotation.

If we now generalize the facts discovered about the electron, we obtain the idea that all elementary corpuscles might well be characterized by a third property—their spin—in addition to their charge and their mass. In the case of the electron the spin is $\frac{1}{2}$—the quantum unit, $\frac{h}{2\pi}$, being assumed. In every complex structure, composed of elementary corpuscles, the spins must be added to each other algebraically in such a way that the entire structure will have a total spin, on which the properties which it manifests externally will in part depend. Applied to the nuclei of atoms, in fact, this idea has proved most fruitful, since it has provided an interpretation of the hyperfine structure of certain spectral lines.

If the complex particles of Matter—such as the nuclei of atoms—have a spin equal to the algebraic sum of the spins of the elementary corpuscles, this spin should be equal to a whole number n, or to half an odd number, $n + \frac{1}{2}$, (in units $\frac{h}{2\pi}$). Briefly, we find that complex particles will have either an even spin (or none), or an odd spin, in units of $\frac{1}{2}(h/2\pi)$, which leads us to divide them into two classes which appear to behave differently with respect to their statistical properties. It is well-known that Boltzmann's classical statistical Mechanics is, from the quantum point of view, no more than an approximation applicable only at sufficiently high temperatures, though in practice it can be applied to a very great number of cases at normal temperatures. Current Quantum Theory, however, has led physicists to substitute, for the enumeration of possible states of a system employed in classical statistical Mechanics, a different method of enumeration, in which two states are considered identical if the only respect in which they differ is the permutation of two elementary corpuscles of identical character. But this new method has been developed in two different ways; either it is assumed that the number of elementary corpuscles in the same state can be as large as we please; or else it is assumed, on the contrary, that the presence of one corpuscle in any given state absolutely excludes the presence of another corpuscle in the same state. In this way we obtain respectively Bose-Einstein, and Fermi-Dirac, statistics. The former is certainly valid for the photon, and leads to Planck's law of spectral distribution; while the second is undeniably valid for electrons, and explains the properties of conducting electrons in metals. The somewhat strange property of electrons, which does not permit a plurality of their number to be in the same state, constitutes Pauli's Exclusion Principle, and, as has been observed, it is the foundation of Fermi-Dirac statistics.

Measurements made of rotational spectra have led to the conclusion that nuclei with an even spin follow Bose statistics, while those with an odd spin (as in the case of electrons) follow Fermi statistics. Scientists have thus been led to believe that the even or odd character of the spin of a given type of material unit corresponds to the validity of the one or the other statistics for assemblages of such units; and this rule was used as a guide by physicists when trying to establish formulae for the structure of atomic nuclei, by introducing protons and electrons as their elementary constituents—and, more recently, neutrons and positive electrons. There is no doubt at all that photons obey Bose statistics, and hence the rule just mentioned leads one to attribute to them an even spin (or none); which again might lead one to believe that the photon is not an elementary unit but consists, for example, of a positive electron and a negative electron. At the same time the fact must be stressed that we have no idea at all as to the origin of the exclusion enunciated by Pauli's Principle, nor of the mechanism which may connect the existence (or non-existence) of this feature with the oddness (or evenness) of the spin, so that it might be rash to seek to generalize a rule inferred from so few individual cases. It is none the less certain that spin is a most important characteristic of corpuscles, and that the laws governing their interaction and their statistical behaviour *en masse* largely depend on it. When experiment reveals the existence of a new corpuscle we must ask, therefore, not only what is its mass and its electric charge, but also what is its spin.

* * *

Let us now return to photons and rays. All physicists will understand that we are here dealing with the various types of radiation of the great electromagnetic series, which ranges from Hertzian waves to γ-rays and to their continuations into the very high frequencies, which characterize exceedingly short-wave radiation and cosmic rays. But on going more deeply into the subject one finds—as usual—that it is no easy matter to give a good definition of what we are seeking to envisage.

The fact is that it is not possible in strictness to speak of waves obeying the classical theories, since the latter take no account whatever of quanta, and assume that energy is uniformly distributed over the wave surface.

With regard to radiation, then, it is natural enough to assume that all photons have the same character; in other words, that the magnitudes which characterize them—such as their proper mass, spin, etc.—are the same for all; their energy, and hence their frequency, alone serving to differentiate them. A consideration of their change of frequency when they are reflected from a rotating mirror seems to prove that *one and the same* photon can pass from one "colour" to another whenever its energy changes.

In the waves of wireless telegraphy, still further, we must assume that the quanta of energy are extremely minute, and that hence there is an enormous number of photons participating in any perceptible manifestation of energy. Hence the continuous and hydrodynamic aspects of such phenomena; and the way in which the existence of photons can be made compatible with the fields considered in elementary electrical theory (like static fields) remains completely mysterious.

The best way to define the different types of radiation might be to consider them as manifestations of quantized energy of such a kind that the frequency ν, experimentally inferred from the quantum $h\nu$, is connected to the wave-length disclosed by optical phenomena through the intermediacy of a velocity not differing by any measurable quantity from the velocity of Light, c. It follows that in the case of scattering, regarded as an impact administered to the photons, we must take (as we do in the theory of the Compton effect) the momentum of a photon as always equal to $\dfrac{h\nu}{c}$.

There is, however, always one disadvantage in this definition, which is that it is based on a mathematical idea, whereas it would be far more satisfactory to base the definition of radiation on some physical property. Further, in proportion as the quantum of radiation increases, new properties appear; a fact which would become extremely important if the principle were to be confirmed that radiation, whose quantum exceeds twice the absolute energy corresponding to the mass of the electron, can become transformed into two electrified corpuscles of opposite sign.

Great stress should also be laid on a phenomenon which has hitherto been observed only in connection with radiation; namely the photo-electric effect, which constitutes a fundamentally important effect of radiation upon Matter; its mechanism, however, still remains completely unknown. We do not know why the energy of the photon is here completely transformed into the kinetic energy of an electron, after first having done the work necessary to detach the latter from the attraction of the nucleus. We cannot repeat too often that a phenomenon of such importance, to which we must attribute the majority of the manifestations of luminous energy, remains completely mysterious, except for the mathematical expression for the exchange of energy given in the Einstein relation.

The difference between the effects, as revealed experimentally, of corpuscles and of photons respectively tends to vanish when the energy of the former and the quantum of the latter become very great; and even the effect of the charge of the corpuscle diminishes, though it does not appear to fall below a certain limit.

Calculation shows that the distinction between certain mechanical properties of rapidly moving corpuscles, and those of photons of very high frequency, also tends to become blurred when the quantum of the photon attains the order of magnitude of the energy of annihilation of the mass of the corpuscle.

*   *   *

What results from all this, in conclusion, is that today the distinction between a corpuscle electrically neutral and having an extremely small mass, and a photon, has become very slight. At the moment it would appear that we have to consider three kinds of neutral corpuscles: the neutron, whose mass is approximately equal to unity; Fermi's neutrino, whose total mass would be very much less than the electronic mass; and the photon, where the connection between the two charges of contrary sign would be such that the mass would be less still.

At present it is believed that a photon may be transformed into two or more corpuscles; it is thought, for example, that the energy of a ray might be capable of giving birth to a couple of electrons of contrary sign, by first providing the amount of energy—$2mc^2$— necessary in order that the masses shall come into existence, and next the kinetic energy which the electrons should possess. The inverse phenomenon again, the dematerialization of a corpuscle, is also believed to be possible and, by extension, the creation of new quantities of Matter at the expense of the kinetic energy of a very rapid corpuscle has also been suggested.

The possibility that two corpuscles which in a certain way are symmetrical—like the positive and the negative electron—can be annihilated, may lead to fresh views about the structure of the photon. A photon consisting of a couple of corpuscles, related to each other in the same way in which the positive electron is related to the negative electron, would then be capable of being destroyed in the presence of Matter by yielding up to it all its energy content. Such an annihilation of the photon, in fact, would constitute the photo-electric effect, and would explain its specific character. At the same time we should understand why a photon, composed of two corpuscles of spin $\frac{1}{2}$, should obey Bose-Einstein statistics.

The idea of the wave and of the energy of a corpuscle have, however, become so fluid that we should experience no surprise to find this new Proteus actually assuming in succession its different aspects under our very eyes. At the same time, if we wish to continue to know exactly what we are saying, we must find a definition of sufficient exactitude for the words we use. But the subtlety, and sometimes the vagueness, of its most fundamental conceptions is one of the difficulties of modern Physics, and not the least among them.

We must not disguise the fact, still further, that these ideas are still largely hypothetical. But experimental investigation, which today is turned on these subjects with the greatest vigour, is certain soon to bring about results capable of introducing some order into the present confusion of our ideas.

2

A DISCUSSION OF THE TWO KINDS OF ELECTRICITY

IF we begin even the most elementary study of electricity we soon read that there are two kinds: that which appears if we rub a piece of glass with cloth, and that which appears if we rub it with a piece of resin; these are the "vitreous" and the "resinous" electricity of the writers of an older generation, known today as positive and as negative electricity. It is a familiar fact that two electricities of the same kind repel, while two electricities of opposite kinds attract one another. No third kind of electricity exists; for example, there is no means of obtaining an electricity which might prove able simultaneously to attract a positively and a negatively charged body. From the shape of our right hand we can pass to that of our left by means of a relation of symmetry conforming to a plan; a fresh symmetry will bring us back to the right hand; but, however long continued, a series of operations of this kind will never give us anything else than the one or the other of these two forms. The alternative between the two kinds of electricity is much the same. If a conductor bearing a positive charge is brought near a second conductor, connected momentarily to earth, a negative charge will be communicated to the second conductor by induction. By means of the second conductor, now that it is charged in this way, we could then, by a similar procedure, give a positive charge to a third conductor, and so on: but this series will never help us to find anything other than these two known kinds of electricity. Clearly this duality of electricity must be something very important and fundamental. So long as the science of electricity developed within the framework of large-scale phenomena and confined itself to the study of bodies—whether bearing electric charges, or magnetized—considered as wholes, and of the electric currents within the conductors similarly regarded, so long no essential dissymmetry between the two kinds of electricity—the two fluids, as the expression went—ever manifested itself. It is true that we customarily assign a direction to the electric current within the conducting wire, and this direction is conventionally that of the movement of the positive electricity within the wire. But this definition is the outcome of a mere convention and does not in the least imply that the positive fluid has a greater mobility; indeed we shall see that in fact it is the negative fluid which was later found to be the more mobile of the two. There is nothing to prevent us from maintaining that there is an absolute symmetry between the two kinds of electricity if once we assume that when a current passes through a wire equal quantities of the two fluids pass within it in opposite directions, the positive electricity moving in the direction which is *conventionally* taken as the positive sense of the current and the negative electricity moving in the opposite direction. This complete reciprocity in the part played by the two kinds of electricity which electrostatics and electrodynamics allowed us to assume to exist, seemed matter of great satisfaction to the minds of scientists—people who are always—but sometimes wrongly—charmed by simplicity and symmetry.

The position was changed completely by the discovery, at the end of last century, of the elementary structure of electricity, an advance which was further intensified at the beginning of the present century. The structure of the two kinds of electricity was shown to be widely different, since positive electricity was linked much more closely than the negative to those properties of inertia in Matter which are symbolized by its mass. The outcome of this contrast is that negative electricity is much more mobile than positive, and consequently plays a much more active part in its motion. This type of preponderance in the case of negative electricity arouses some regret that it should be this kind for which "negative" has been reserved by convention, and with this the minus sign in equations. Today, however, the convention has been sanctioned by use to a degree which forbids any change;

and this breach of the symmetry between the two kinds of electricity has been of too great an importance in the development of contemporary Physics not to compel us to spend a few moments upon it.

* * *

The decisive factor in this evolution was essentially the discovery of the electron. I do not wish to detail here its various stages, with which the names of Crookes, J. J. Thomson, Lenard, Villard and Jean Perrin are associated. In greatly varying circumstances different scientists succeeded in demonstrating the existence of minute electrified particles, each of them similar to all the others, and emerging from the very depths of Matter. Sometimes these particles spring from the cathode of a high vacuum tube (cathode rays in a Crookes tube), at others they are spontaneously projected by radioactive substances during transmutation. At other times again they can be observed rising from a substance illuminated by certain rays (photo-electric effect) while at others they escape from Matter in a state of incandescence (thermionic emission). These particles are always of the same kind, and their mass and charge are also always the same; whence the fact was grasped that they are one of the essential materials—one of the elementary bricks— of which Matter is built up. Experiments further showed that these particles bear a negative electric charge, and it was therefore assumed that negative electricity had a discontinuous corpuscular structure and was divided into corpuscles all of which were similar to each other. The name of electron was henceforward confined to these corpuscles, every further advance made by Physics tending to confirm the existence, and the identity, of these electrons, and to show that every fragment of Matter contains a vast number of them within its internal structure. The discovery of the Zeeman effect—i.e. the modification which spectral lines emitted by a given source undergo when the source is placed within a uniform magnetic field—brought powerful support to this view, by showing that the entities which cause this emission from within the atom are precisely those negatively charged particles which had been found outside Matter in the experiments mentioned above.

The discontinuous structure of negative electricity was thus founded on a firm basis. The electron was the element of the structure, and it soon became familiar to physicists, who learned to measure its charge and mass despite their extreme minuteness, and to study its motion under the influence of electric and magnetic fields. Research of this kind soon showed that it sufficed, in order to represent the properties of the electron, to treat it as a minute corpuscle of given mass and charge obeying the laws of Mechanics. We shall shortly recall the fact that this view of the electron, although adequate as a first approximation, had to be modified later. The important question with which scientists were faced from the beginning, however, was this: given the discontinuous structure of negative electricity, is it paralleled by a discontinuous structure of positive electricity?

As soon as this question was clearly stated, it was seen that even if positive electricity had a discontinuous structure, this structure was certainly not rigorously symmetrical with that of negative electricity. Thus, while it had been relatively very easy to cause electrons to be ejected from Matter, it is a far less simple affair to detach a positive charge from it. Further, whenever particles with a positive charge are obtained—e.g. those constituting the a-rays of radioactive substances—it is always found that these have a mass which, though extremely minute actually, is very much greater than that of the electron. It may be said, therefore, that the properties of inertia and weight in Matter appear to be much more closely connected to positive than to negative electricity. Thus the lightest particle of positive electricity which it had been possible to identify (until 1932) is nearly 2,000 times heavier than the electron.

This difference in the parts played by the two kinds of electricity became gradually clearer in proportion as our knowledge of the structure of Matter became wider and sounder. Matter, needless to say, has an atomic structure. The success of atomic theories in Chemistry, and of kinetic theories in Physics, had made

the assertion probable, to begin with, and it was definitely confirmed by the celebrated experiments associated with the name
of Jean Perrin. But once the atomic structure had been accepted,
everything suggested that the atoms of the elements were themselves complex structures, at whose interior a more profound
analysis would once again find electrons. Otherwise, how explain
the fact that Matter behaves as a source of electrons in so many
different circumstances? Without following in detail the history of
the attempts made to construct an electric model of the atom, I
shall state directly the solution to which the progress made about
1910 led. The memorable experiments made by Lord Rutherford
and his collaborators on the scattering of α-rays by Matter have
furnished a proof that the atom, so far from being comparable to
a completely filled sphere, as the success of certain arguments
(somewhat over-simple) used in the kinetic theory of gases might
have suggested, was almost completely empty, most of its Matter
and all its positive electricity being concentrated in an extremely
minute central region. The sphere of influence of the atom, as
defined by the kinetic theory of gases, has a radius of the order of
1/10,000,000th of a millimetre, while the central region I have just
mentioned has a radius about 100,000 times smaller still. If we
remember that volumes vary as the cube of the linear dimensions,
we shall see how extremely minute is the position occupied in the
atomic region by this nucleus where the positive charge resides.
Once Lord Rutherford had discovered how very intensely the
positive charge is concentrated at the centre of the atom, he
suggested a model for the atom which under the name of the
planetary model met with great success. According to this idea
the atom was to be compared to a miniature solar system formed of
electrons which, taking the part of planets, gravitated under the
action of Coulomb forces around the central sun bearing a positive
charge. Every simple element was to be characterized by the
number N of its electrons (planets); and, since the atoms in their
normal state are certainly electrically neutral systems, the central
sun must have a positive charge equal to N times the charge of the
electron, and of contrary sign, given that the atom contains a certain

number N of electrons. Hence the nuclei of the different kinds of atoms bear electric charges which are integral multiples of an elementary unit equal to the charge of the electron and of opposite sign. The simplest of these nuclei would then be the one bearing only one elementary unit, and experiments have proved that this is the nucleus of the lightest element, Hydrogen. Thus the Hydrogen nucleus plays the part of the unit of positive electricity, and the name of proton has been set aside for it. Now the proton has a mass some 2,000 times greater than that of the electron. The Hydrogen atom, the lightest known atom, would then, according to Rutherford's theory, consist of a proton with a charge e, containing practically the whole of the mass of the atom, and of an electron (playing the part of the planet) having a charge of $-e$ and revolving around the proton. The latter, on account of its great mass, could be regarded as practically at rest, and as not reacting to the influence exerted on it by the planetary electron. In the case of the atoms of the heavier elements the nucleus has a higher charge, and its mass is greater. It did not take long to discover that the number N, which characterizes each kind of atom by fixing the composition of its attendant train of electron-planets, coincides with the number denoting the position of the element in the Periodic Scale of the elements, when they are arranged in ascending order of atomic weights, certain relatively unimportant inversions within the Scale only being introduced. Thus the atom bearing the atomic number N has a nucleus with N units of positive charge; it is not, however, formed of N protons, as is shown by a comparison of its charge with its mass, and it was thought during the early part of this planetary theory that it consisted of a firmly combined block of protons and electrons, the number of protons exceeding by N that of the electrons, as demanded by the total charge. It may be added that nothing prevents us from assuming that one and the same total charge of the nucleus can be brought about by two or more combinations of protons and electrons, since it is only the difference between the respective numbers of these constituent entities which is fixed by the total charge. We thus come to envisage the possibility that certain atoms may have the same number N, and hence may possess in bulk the same chemical properties, but may have a different mass—i.e. elements practically identical as far as their properties are concerned but having different atomic weights. Later the discovery of isotopes came to show that this potentiality is in fact realized in Nature.

Accordingly—and this is the essential principle at this point—practically the whole of the mass of the atom is concentrated in the nucleus, where it is invariably associated with a positive electric charge. Around this nucleus negative electricity is distributed in the form of a moving cloud of electrons, whose mass is practically negligible. To understand the degree to which this atomic structure is the expression of a deep asymmetry between the parts played by the two kinds of electricity, it is enough to observe that a converse structure, with the negative electricity concentrated in a relatively heavy nucleus surrounded by much lighter positive charges, is quite as probable a priori. But experiments show in the clearest possible manner that this inverted system does not in actual fact exist: the real, physical, atoms have a positive central charge, and this fact clearly illustrates the specific affinity existing between mass and positive electricity; an affinity whose result is to destroy completely any symmetry between the two kinds of electricity.

I need not here, however, explain the development of the planetary theory of the atom, nor how, thanks to Bohr, it was enabled to take on an exact quantitative form, the success of which is by now quite familiar. The essential characteristic of the progress effected by Bohr is that he introduced the principles of Quantum Theory into the planetary model of the atom. Later still, Wave Mechanics came and modified the formulation and the interpretation of these principles, at the same time transforming considerably our conception of electrons as well as of the elementary corpuscles of Physics in general. According to these new ideas, then, these elementary corpuscles are not adequately to be rendered by the concept of exactly localized material points—an idea which is, indeed, implied in the very use of the word "corpuscle." The concept is certainly useful for describing certain aspects of the elementary processes, but, in order to describe certain other aspects, we must add to it the conception of waves, so that a complete description of the electron, for example, makes use of both concepts simultaneously. The result is to transform the theory of the planetary-atom, since it is the interference of the waves associated with the electrons that determines the existence of the stationary states of the atom. All these changes, however, interesting as they are in themselves, and important as is the part they have played in the history of contemporary Physics, do not affect the foregoing presentation of the points with which the present discussion is concerned. Undeniably, the changes mentioned have profoundly modified the description of the elementary units of electricity; but the fact still remains that the elementary units of positive electricity are always associated with masses which are very much greater than those of the units of negative electricity.

Actually, the position is that two discoveries of the utmost importance, made as a result of experiments in 1931 and 1932, have given us valuable new information on the elementary structure of Matter, and have opened unexpected horizons regarding the asymmetry between the two kinds of electricity. What I have in mind is the discovery of the neutron and of the positive electron. These are now to be dealt with.

* * *

I have mentioned that it had long been assumed that the nuclei of atoms are complex, consisting of protons and electrons. This idea had been suggested by the instinctive belief of physicists in the unity of Matter, and it had become practically necessary to adopt it after the existence of radioactive phenomena had been made familiar by the admirable work of Henri Becquerel and of M. and Madame Curie. As I have already observed, the radioactive substances are heavy elements, and bear the highest numbers in the Periodic Table of elements.[1] The discovery of radioactivity

[1] cf. p. 31, for additional details.

was of so great interest to physicists because it proved to them that atomic nuclei are complex structures capable of breaking down into simpler structures; it also showed them how electrons or light nuclei emerged from the depths of a heavy nucleus, a fact which confirms the existence of electrons and of positive particles at the interior of nuclei.

Research into artificial disintegrations has in recent years inspired a very considerable number of investigations and has led to extremely interesting discoveries. I cannot here go into these investigations and discoveries, of which even a cursory examination would take us very far from our subject: yet the fact ought to be mentioned that M. and Madame Joliot and Prof. Chadwick, in the course of certain research work on artificial disintegration, discovered independently that among the products of disintegration there were corpuscles of a character hitherto quite unknown. These corpuscles pass very easily through Matter—they are, to use the physicists' term, very penetrating; they have no electric charge and possess a mass very near that of the proton. These are neutrons, and it seems certain that these neutrons must exist at the interior of the atomic nuclei, playing an essential part, and one unsuspected only a few years ago, within their structure. So true is it that our theoretical notions are essentially provisional and always at the mercy of some new discovery!

Within less than a year of the discovery of the neutron a fourth kind of elementary corpuscle was found. In the course of research on the effects of disintegration caused by cosmic rays—those mysterious rays which reach us from the depths of inter-stellar space—Anderson, and Blackett and Occhialini, independently demonstrated the existence of positive electrons, i.e. of corpuscles which very probably possess the same mass as negative electrons, having the same charge but of contrary sign. These positive electrons (or positrons) whose existence Dirac had predicted before their discovery, on the strength of bold theoretical arguments, appear to be unstable in the sense that, when they penetrate Matter and there meet with negative electrons, they are liable to disappear and, while disappearing, to neutralize and consequently to annihilate a negative electron. Dirac's theoretical arguments (to repeat) had led this fact to be predicted; and the brilliant experiments carried out by Thibaud and Joliot seem to have established it firmly.

In consequence of the sensational discovery of the neutron and the positive electron the position has been greatly complicated: We must now ask exactly what position we have reached with respect to the asymmetry between the two kinds of electricity.

*   *   *

Today we are acquainted with four types of corpuscle which we are inclined to regard as elementary. In fact, however, the proton and the neutron are sometimes considered to be, in reality, two different states of one and the same particle. In that case, the change from proton to neutron occurs together with the creation and emission of a positive electron, while the converse transition from neutron to proton, on the other hand, is associated with the creation and emission of a negative electron. This hypothesis, therefore, involves the actual existence of only a single elementary particle of relatively heavy mass—the neutron-proton—which may then assume two contrasted forms, one of which is electrically neutral and the other electrically positive; and when introduced into recent theories of the atomic nucleus the hypothesis enables us, more specifically, to regard the interactions between the nuclear components as "exchange interactions," as would seem to be necessary in order to explain the stability of nuclear structure. This conception is also involved in Fermi's brilliant and profound theory of the continuous spectra emitted by radioactive substances.

But whether or not we adopt the hypothesis of the neutron-proton, we are in any case led to regard the two electrons as simple.[1]

---

[1] Recent research seems to indicate the existence of two other types of corpuscle, thus making six in all:—the neutrino, which appears to have a vanishingly small mass, or perhaps no mass whatever (cf. p. 141), and also the heavy electron, sometimes called the barytron or mesotron, with probably a mass of from 150–300 times the mass of the electron, which is here taken as the unit, and with an electric charge nearly equal to that of the electron. The properties of these new types of corpuscle, however, are not yet at all well known.

This suffices to restore a certain degree of symmetry between the two kinds of electricity; for the two electrons, whose masses are doubtless identical, while their charges are equal and of contrary sign, appear to be quite comparable with one another. A complete theory of the two electrons—the suggestions advanced by Dirac give us at any rate a preliminary idea of it—will probably succeed some day in revealing the real character of the symmetry between them, a symmetry which, to revert to a comparison made at the beginning of this Chapter, is doubtless analagous to that between the right and the left hand. One fundamental difference, however, does exist between them. For the negative electron is constantly manifesting itself in our experiments, whereas the positive electron only makes an exceptional appearance, and always has a tendency to disappear when in contact with Matter. I shall revert to this point.

But the symmetry between the two electrons appears in very different lights according to whether we assume that the proton is elementary or not. If it is—whether the neutron is so also is of small importance—then the existence of the proton maintains a complete asymmetry between the two electricities, since positive electricity can exist under two corpuscular forms which are irreducible; one of them, the positive electron, being completely symmetrical with the element of negative electricity, while the other, the proton, has no known negative counterpart. If we adopt this hypothesis of the simplicity of the proton, then the discovery of the positive electron does not appear to have removed the structural asymmetry between the two kinds of electricity to which the evolution of atomic Physics had led. But the case is different if it is assumed that the neutron is simple and the proton complex. On this hypothesis, which is certainly more intellectually attractive than the preceding, the electrons would be elementary and completely symmetrical units of the two kinds of electricity. They would have equal and very minute masses, and the real unit of heavy Matter would then be the neutron, an element without charge but at the same time the site of practically the entire mass of material substances. Finally, however, the part played by the two kinds of electricity and their relation to mass is asymmetrical: this much is definitely established by the totality of our knowledge of the atomic world. The asymmetry, therefore, must necessarily reappear somewhere, and if it is eliminated from the structure of the two kinds of electricity, it must reappear in their properties. Actually, the assumption that the proton is complex compels us to assume that there is an affinity between the positive electron and the neutron, by which the very frequent occurrence of the combination known as the proton is explained—since the proton is one of the essential elements of material structures. Certain authors have maintained, however, that it is not impossible for the negative electron to unite with a neutron to produce a "negative proton"; but this combination, if it exists at all, is certainly much less frequent than that which would give the positive proton on our present hypothesis. Accordingly, if the proton is complex, there is certainly an affinity, or tendency towards combination, which is much greater between the neutron and the positive electron than between the neutron and the negative electron. This is the natural explanation of why the positive electron is much less frequently observed than the negative electron; for under the influence of the affinity of the neutrons contained in Matter, it would be continually in the state of combination forming the proton, and would appear in a free state much more rarely than the electron, whose affinity with the neutron is much weaker, if indeed it exists at all.

It follows that the recent discoveries of the neutron and of the positive electron do not help us to remove the asymmetry between the two kinds of electricity. If we assume the complexity of the proton we succeed in establishing a structural symmetry between them; but in that case we find the asymmetry reappearing in their properties—in the fact that positive electricity has a very much greater tendency to be associated with mass; and it does not look as though this asymmetry could ever be avoided, since it is a fact supported by the entire body of our experimental investigations in the atomic domain during the last 40 years. Nevertheless there is a certain satisfaction for the mind in seeing some possibility of restoring a complete structural symmetry between the two kinds of electricity, and in transferring the problem of asymmetry to the question of the relation between mass and electricity. Formulated in this way, the problem is one of great interest; it is very far from being solved, and it serves as a measure of the full extent of our ignorance. At bottom, we have no exact knowledge of the relation between mass and electricity. At one time it had been thought that we could maintain that all mass was of electric origin, an assertion resting on the fact that the mass of elementary corpuscles varies as a function of velocity, in accordance with Lorentz's law. The development of the Theory of Relativity, however, had gravely shaken this argument, since it had shown that all mass, whatever its origin, must vary with velocity in accordance with Lorentz's law. The discovery of the neutron, still further, hardly seems compatible with the view of the purely electrical origin of mass. The attempt to reduce mass to electricity having thus failed, we must revert to the view which holds these two entities to be distinct, and must seek to define their relation to each other. The theory of this relation, if we ever succeed in forming one, will first of all have to explain why there is a special affinity between mass and positive electricity. The affinity is of capital importance, since upon it rests the entire structure of the physical world. No doubt the solution of the problem lies in a distant future: nevertheless the discovery of the neutron and of the positive electron has provided us with some essential data on the subject which we were wanting until then. The question is whether these data will be sufficient to allow us to reach the goal: but to risk an answer would be rash. It is the secret of the future.

3

THE EVOLUTION OF THE CONCEPT
OF THE ELECTRON [1]

IT is the aim of the present Chapter to trace the evolution of the idea of the electron, and more generally that of the elementary corpuscle, during the last 50 years. Again and again in the course of this account we shall see how greatly the fundamental notions of the science of electricity—notions which were successively acquired at the beginning of last century by the labours of the pioneers in that science—play an essential part in modern notions about the constitution of Matter and the structure of the elementary entities composing it. In this way, by the mere statement of the facts, we shall be associating ourselves with the tribute rendered to the illustrious memory of André-Marie Ampère; for without him there would have been no electromagnetic theory nor a theory of the electron; while all the splendid progress of contemporary Physics would have been rendered impossible, or at any rate postponed.

* * *

To compare the flow of electricity along a conductor with that of a liquid in a pipe has become familiar and even slightly commonplace. This comparison, which is often found helpful towards an understanding of the laws governing electric currents, almost inevitably suggests the question whether the electric fluid has a continuous or a discontinuous structure. Actually, the success of the Atomic Theory of Matter, supported by numerous experimental proofs, has shown that the material fluids with which we

[1] Lecture delivered at Lyons, March 7, 1936, on the occasion of the Centenary of the death of André-Marie Ampère.

compare the electric fluid have a discontinuous structure; and the question arises naturally whether there are such things as elementary corpuscles of electricity. The question becomes even more reasonable as soon as we consider the laws of electrolysis known since the time of Faraday. Interpreted in the light of the Atomic Theory of Matter, these laws mean in effect that every ion always carries with it an electric charge equal to an integral multiple of the elementary charge. This fact not only implies that there is such a thing as an elementary electric charge, but it also allows its magnitude to be calculated as soon as we have obtained Avogadro's Number by some other method; Avogadro's Number being the number, the same for all substances, of the molecules contained within one gramme-molecule.[1] Nevertheless this proof of a discontinuity in the structure of electricity, due as it was to the laws of electrolysis, was somewhat indirect, and for a long time our knowledge of Avogadro's Number (of which the Kinetic Theory had yielded certain indications) was insufficient to allow us to calculate exactly the value of the elementary charge. The calculation did not really become feasible, in fact, until the time of Jean Perrin's celebrated experiments. In order to provide a genuine proof of the discontinuous nature of electricity, it was essential to isolate and study the elementary corpuscle of electricity; and this was effected by a brilliant series of experiments towards the end of the nineteenth century, at any rate so far as negative electricity is concerned.

I cannot here recall in detail the stages of this "conquest of the electron," with which the names of J. J. Thomson, Lenard, Villard and Jean Perrin—to mention only those which first occur to my own mind—are associated. Gradually it came to be seen more and more clearly that, under certain conditions, material substances are capable of expelling corpuscles of electricity, which thereupon travel freely until they are again buried within the depths of Matter. Sometimes, as I have remarked in an earlier Chapter, they appear in discharge tubes (Crookes tubes) in the form of

[1] [The gramme-molecule is the number of grammes equal to the molecular weight of any given substance; Avogadro's Number is $6 \cdot 16 \times 10^{23}$.]

F

cathode rays, while at others they have been discovered among the products of disintegration of radioactive substances in the forms of β-rays. At other times, again, they are found emerging from substances raised to a high temperature or from those exposed to Light or X-rays, in which case we speak of the thermionic effect or of the photo-electric effect. The essential point, however, is that the corpuscles of electricity—the electrons—obtained by all these different methods, are invariably found to be identical with each other. They all bear the same quantity of negative electricity, and their properties, which are well defined, are always the same.

The properties of the electron which were ascertained during the first years after its discovery could be summed up by saying that the electron always behaved as though it were a corpuscle of negligible dimensions, possessing mass and an electric charge, both of them very minute but well defined. This implies that the electron indicates its passage by effects which are strictly localized at a point in space—at any rate within the range of human observation—and further that, in either the presence or the absence of electromagnetic fields, it always moves in the way in which according to classical Mechanics a corpuscle (regarded as a point) would move, having a very minute but clearly defined mass and bearing a small and unvarying negative charge. The deviation of the path of the electron, in electric and magnetic fields of known intensity, enables the ratio between its charge and its mass to be measured. The knowledge of Avogadro's Number, again, allows its charge to be ascertained, while Millikan's brilliant experiments permit us to obtain this directly. In this way, then, the magnitude of the charge and mass of the electron—both of which are extremely minute—can be determined.

Further investigation of the mechanical properties of electrons moving with high velocity has since shown that the mass of the electron varies with its velocity, in exact accord with the law predicted by the Theory of Relativity. We know, still further, that the development of Einstein's brilliant ideas on the relativity of physical phenomena leads to the adoption of a dynamics differing from the classical Newtonian dynamics when we are dealing with high velocities—i.e. those approximating to the velocity of Light in empty space. We can formulate this difference by saying that the mass of a corpuscle, instead of always remaining constant, as Newtonian dynamics assumed, increases substantially with its velocity when the latter approaches that of Light, and even tends' towards infinity when the velocity approximates still more closely to that of Light, which shows that if Einstein's conceptions are correct, no fragment of Matter can ever move as fast as, or a *fortiori* faster than, a ray in empty space. A lengthy series of experiments, of which those conducted by Guye and Lavanchy are the most famous, have proved that this variation of mass with velocity does in fact exist for electrons moving with high velocity. This noteworthy confirmation of relativistic ideas, however, while of great importance from the standpoint of general theoretical Physics, has not added anything essentially novel to the concept of the electron. For after the confirmation, as well as before, the electron had to be regarded as a corpuscle resembling a geometric point, or at any rate of very minute dimensions, having a certain charge of negative electricity and possessing a certain exactly determined mass when at rest. The only new fact is that the apparent mass increases whenever the electron is put into rapid motion. But the growth of our knowledge has since led to much greater changes in our views of the electron, as we shall see below.

The part played by the electron, then, is that of the elementary corpuscle of negative electricity; and the question now is whether there are corpuscles of positive electricity. Some time passed after the discovery of the electron before we possessed any definite proof of this. Later, however, it became certain that positive electricity too is divided into corpuscles; and the elementary corpuscle of positive electricity has now been identified with the proton—the nucleus of the Hydrogen atom. Today, when the discovery of the positive electron is still of quite recent date, we must ask whether the proton can fairly claim to be the elementary corpuscle of positive electricity. Previously no doubt had been felt for many years about the matter, and for the time being we shall adhere to this view.

Research into the properties of the proton led physicists to treat it as a corpuscle of the nature of a geometric point, or at any rate of very minute dimensions, and analogous to the electron. At the same time this corpuscle has an electric charge which has turned out to be positive and exactly equal in value to that of the electron. Its mass when at rest, however, is very much greater than that of the electron at rest: it is about 2,000 times greater, and this difference establishes a curious asymmetry between the two kinds of electricity.

To sum up, experimental research before 1920 allowed us to regard electricity as consisting of elementary corpuscles somewhat of the nature of geometric points, having an exactly ascertained electric charge and mass when at rest. I shall now explain the way in which physicists tried to picture the structure of Matter on the hypothesis that it consists of aggregates of elementary corpuscles of electricity; and how this attempt, by leading us into the internal structure of Matter, caused us to see that the corpuscles of electricity had very curious properties—properties differing widely from those which might have been expected in a simple corpuscle approximating very closely to a geometric point. In this way we shall be all the readier to understand the recent development of our concept of elementary corpuscles.

*   *   *

Once experiments had begun to indicate that the electric fluid was of a discontinuous nature, the theorists took hold of this idea and tried to base explanations on it. Sir J. J. Thomson was among the pioneers in this direction; but it was H. A. Lorentz who more particularly tried to reconstruct the whole of electromagnetic Theory by the systematic introduction of electrons. I shall not dwell here on all the successes which Lorentz's Theory enjoyed. The most brilliant of these was to predict the triplets and doublets found in the normal Zeeman effect, whose experimental discovery in 1896 confirmed Lorentz's ideas, proved that Matter contains negative electrons, and even allowed a first estimate to be made of the ratio between their charge and their mass. But the contribution of the theory of electrons went further: it allowed the laws of dispersion and scattering to be traced in this field, and it permitted a large number of electro-optic and magneto-optic phenomena to be predicted, and various electric and thermal properties of metals to be interpreted, etc. The theory of the acceleration-wave, again, seemed to show us clearly how radiation originates from the motion of the electric charges contained in Matter.

Certain shadows, however, soon came to darken the brilliant successes of electron theory. The most disturbing of these arose from the current theory of radiation in thermal equilibrium. The theory of electrons enabled us to calculate what should be the distribution of energy between the frequencies in the radiation which normally exists within an enclosure maintained at a constant and uniform temperature. If the way in which the exchange of energy between Matter and radiation proceeds within the enclosure in question is analysed by the help of the electron theory, we succeed in finding a certain Law of spectral distribution of the energy of the radiation in a state of equilibrium. Unfortunately this Law—that of Rayleigh and Jeans—does not agree with experiment; and while it adequately represents the facts in the region of low frequencies and high temperatures, it is seriously incorrect for high frequencies and low temperatures. This was a serious check to the electron theory, for the Rayleigh-Jeans law was the necessary consequence of the body of ideas held at the time regarding the wave nature of radiation and the discontinuous structure of Matter and Electricity. To obtain a formula different from that enunciated by Rayleigh and Jeans, and correctly representing the result of experimental research, Planck was compelled to introduce his famous Quantum Theory, according to which the electrons contained within Matter can only assume certain states of motion. These special states of motion—quantized states—were ascertained by Planck from rules into whose formulation he introduced the famous universal constant having the dimensions of action, and since then known as Planck's Constant. I cannot here outline the Quantum Theory such as it became in consequence of Planck's work; at the same time I must stress the following—the essential—idea. By showing that the electrons contained within Matter behave in a way very different from that of the material points assumed by classical Mechanics—or even by relativistic Mechanics—the Quantum Theory showed from its very beginning that, in order to represent the totality of the electron's properties, it would not do to remain satisfied with the rather over-simplified simile of a corpuscle of the nature of a geometric point, or something approaching to this, whose sole characteristics were its mass and its electric charge. The conditions of quantum stability imposed by Planck and his successors on the small-scale motion of the electron cause the electron's entire trajectory to appear as though the electron itself were present simultaneously at every point of this trajectory. It was only gradually, in fact, that the implications of this important fact were realized by investigators, and for a long time Quantum physicists were satisfied to employ simultaneously the simile of the electron, regarded as a material point, and the conditions of quantum stability which, however, involved by implication the inadequacy of that very simile. This simultaneous use of somewhat contradictory concepts was the basis from which our knowledge of the structure of the atom began to grow, and more particularly the famous Theory associated with the name of Bohr.

Thus during the first few years of the twentieth century physicists, having proved that material substances consist of atoms, began to try to form some idea of the structure of the atoms of the various simple substances. The ease with which it had proved possible to detach numbers of electrons from Matter in greatly varying circumstances induced the belief that the electron must be one of the essential constituents of atomic architecture. Certain attempts had already been made to obtain what might be called a model by which the properties of the atom might be represented, when the memorable experiments made by Lord Rutherford and his collaborators on the deviation of α-rays during their passage through Matter provided a proof that a nucleus bearing a positive charge, and having extremely minute dimensions relatively to the entire atomic structure, exists at the centre of the atom. Lord Rutherford, reverting to a suggestion formerly made by Jean Perrin, now proposed that atoms should be compared to miniature solar systems, in which the central sun possessed a charge of positive electricity equal and of opposite sign to an integral multiple of the charge of the electron. Around this "sun" the electron-planets were supposed to revolve, and under normal conditions the resulting system would be electrically neutral. Passing from one chemical element to the next one would find the nuclear charge increasing by one, the number of the electrons revolving around it also growing by one. This atomic model certainly afforded a satisfactory explanation of some characteristics of actual atoms; but it also met with serious difficulties. For since the intra-atomic electrons described (as it were) planetary orbits around the central positive "sun," under the action of Coulomb forces, they ought to have had the capacity, according to classical Mechanics, of executing an infinity of different motions. Further, according to the general results of electron theory, they should radiate energy continuously in the form of radiation having continuously variable frequencies. In that case, however, atoms would be unstable, and Matter would rapidly be annihilated; further, there would be no explanation of the discontinuous character and of the unvarying structure of the spectra emitted by the elements. To remove these difficulties Bohr, in 1913, evolved the striking notion of applying to the Rutherford model the new principles of Quantum Theory, instead of the classical laws of Mechanics and electro-magnetism. He assumed, therefore, that the planetary electrons could describe only a limited number of the orbits that were potentially open to them under the classical laws—precisely those orbits, in fact, which satisfy the criteria of quantization already formulated and employed by Planck. He further assumed that the planetary electrons, in moving in their quantized orbits, emitted no radiation, which was in absolute contradiction with the classical theory of the acceleration-wave. Finally, Bohr advanced the hypothesis that electrons are capable of suddenly changing their orbits, at the same time emitting part of their energy in the form of radiation, the frequency of this being calculable by dividing the energy lost by the electron by Planck's constant. On such a foundation he built a detailed theory, the striking success of which is familiar to all who have followed the development of contemporary Physics. This theory not only explains the stability of the atom and the permanent character of spectra, but it also enables us to predict exactly the structure of optical spectra and of the spectra of X-rays, to work out numerically Rydberg's constant, and even to predict the extremely slight variation of this constant when we pass from Hydrogen to Helium, etc. The essential postulates of Bohr's Theory have been confirmed by research on the phenomena of ionization by collision, and it has rendered possible a preliminary explanation of the chemical properties of the elements and of the periodicity of these properties —a periodicity readily perceived on reading the list of elements arranged in the order of their increasing atomic weights.

The Quantum Theory of the atom, perfected in 1916 by Sommerfeld, who introduced into it relativistic dynamics in place of classical dynamics, by which substitution he was enabled to give a more detailed account of the structure of certain spectra, was in a position correctly to predict the normal Zeeman effect (which had already been interpreted by Lorentz's theory of electrons) and also the Stark effect, of which there was no completely satisfactory interpretation. Thus the whole system, known today under the name of "the old Quantum Theory," developed on the basis of Bohr's ideas. It was a theory successful in many directions, and it rendered inestimable services to atomic Physics by providing it for the first time with a systematic schema. From the point of view which interests us at the present stage, however, the characteristic feature of the old Quantum Theory is that it finds room side by side for the concept of the electron regarded as a corpuscle and obeying the laws of dynamics, and for the new ideas of the Quantum Theory —a somewhat illogical proceeding. Now these new ideas, as I have already observed, lead us to regard quantized orbits as being dynamic units which must be envisaged in their totality; and this is incompatible with the classical view of an orbit described progressively by a corpuscle taken as a geometric point. Further, by introducing integers into the formulae of quantization, the Quantum Theory compelled us to bring in an element completely incompatible with the structure of the old dynamics, even as amended by the Theory of Relativity; for this structure is essentially continuous. Thus the old Quantum Theory was a compromise, and for this reason could not be regarded as satisfactory. It was felt that a more coherent edifice would have to be built up; and from 1923 onwards the appearance of the New Mechanics did much to improve the position; but in the process—as we shall see—a profound modification of our view of the electron had to be introduced.

Before speaking of Wave Mechanics, however, I must first show that it no longer seemed feasible to describe the electron solely in terms of its mass and electric charge, even before the beginnings of Wave Mechanics, and even while it continued to be regarded as a simple corpuscle. Numerous and important experimental facts had already compelled physicists to attribute to it an "internal" rotation and a magnetism of its own, and this complication would have been enough by itself, and apart from the difficulties previously mentioned, to show that our original view of the elementary unit of negative electricity was an over-simplification.

*   *   *

It is known that the light-rays sent out by a given source are modified if this source is placed within a sufficiently intense magnetic field. This is the Zeeman effect—the discovery made by the famous Dutch physicist in 1896. One of the great advantages of the Electron Theory had been that it had enabled Lorentz to predict exactly the phenomenon discovered by his compatriot. Further investigation, however, soon showed that the Zeeman effect, as predicted by Lorentz and actually observed by Zeeman, is really quite exceptional. Only when the light-source consists of certain specific substances can the simple modifications predicted by Lorentz's theory be observed with certain rays; generally, the modifications produced in the rays by the presence of a magnetic field are much more complicated than the theory of electrons predicted. This fact is formulated by saying that abnormal Zeeman effects are a great deal more frequent than the normal Zeeman effect, and that it was a lucky accident which caused Zeeman at the very beginning of his work to come across instances in which the normal effect occurs. Naturally enough Lorentz, and those who continued his researches, tried to extend the original theory of the Zeeman effect so as to make it include abnormal effects. But the attempt did not succeed; and when the Quantum Theory of the atom was formulated, thanks to Bohr, it might have been hoped that this new road would lead to a theory of the Zeeman effect embracing these abnormalities. But here too there was a fresh disappointment. The old Quantum Theory, applied to the influence exerted by a magnetic field upon the spectrum emitted by an atom, led to the discovery of the precise results found by Lorentz in the case of the triplets and doublets of the normal effect; but it remained wholly impossible to interpret the anomalies. The inadequacy of our theories of the electron was thus definitely revealed.

Abnormal phenomena of the same type appeared later also, when the development of the Quantum Theory of the atom enabled the optical and the Röntgen spectra of the elements to be predicted and analysed exactly. Bohr's theory had permitted the composition of the spectral series to be interpreted to what might have been called a first approximation. Then relativistic dynamics added certain corrections, and by allowing for these Sommerfeld obtained a more exact approximation which enabled him to predict the fine structure of spectra in greater detail. But this further approximation in its turn was found to be inadequate; the actual structure of spectra is found to be still more complicated, when its details are scrutinized, than Sommerfeld's theory predicts. At this point it was realized that the Quantum Theory, even in the extended form reached after the introduction of the corrections due to the Theory of Relativity, could not account fully for all the complexities of the spectral series. Quite obviously some indispensable element was lacking.

Consideration of these difficulties led Uhlenbeck and Goudsmit to grasp, to their great credit, that the dilemma was due to the over-simplification of the concept of the electron which formed the foundation on which the development of the Quantum Theory of the atom rested. Their suggestion was, therefore, that the electron should no longer be regarded simply as an electric charge but also as a miniature magnet, so that, in addition to its charge, the electron would possess a certain magnetic moment. It would also have a certain "internal" angular momentum, analogous to the angular momentum of a solid body, rotating on an axis. To give a visual representation of these new properties of the electron, regarded as a corpuscle, Uhlenbeck and Goudsmit represented it as being a minute sphere of negative electricity rotating on one of its diameters, the rotation generating both intrinsic angular momentum and magnetic moment. These two were thus closely interconnected. Next, guided by considerations which cannot here be set out in detail, they gave greater exactitude to their hypothesis by attributing, to the "internal" magnetic moment and angular momentum, values which were exactly calculated and expressed by magnitudes frequently met with in Quantum Theory. According to these views the "internal" motion of the electron is quantized, the corresponding angular momentum being equal to half the usual quantum unit of angular momentum.

The introduction of this group of supplementary hypotheses regarding the electron into the Quantum Theory of the atom enabled Uhlenbeck and Goudsmit to show that the anomalies of the Zeeman effect, the fine structure of optical and of Röntgen spectra respectively, and other difficult phenomena to which I could not refer previously, and which are known as gyromagnetic anomalies, could now be explained. These results, confirmed by those obtained by other investigators, demonstrated clearly that our view of the electron must inevitably be completed in the sense indicated by the two Dutch physicists.

Thus even if the difficulties connected with the interpretation of quanta are set aside—and these difficulties seemed to make it necessary to abandon, at least partially, the corpuscular character

of the electron—the simple assimilation of the electron to a material
point bearing an electric charge undeniably appeared to be inadequate. The electron has a kind of "internal motion" having an
axial symmetry—a quantized motion that is inseparable from its
very existence; this new fundamental characteristic of the electron
is its spin.[1] In addition to its mass and its electric charge, therefore,
the electron has a third and equally fundamental property, its
spin, which must be regarded from both its kinetic and its magnetic
aspect. No complete theory of the unit of negative electricity, then,
can ignore it. I shall, nevertheless, disregard it for a few moments
in order to set out the development of Wave Mechanics in its
original form.

*   *   *

The origin of Wave Mechanics cannot be properly understood
without a brief survey of the development of the Theory of Light
during the last thirty years. The essential fact of this development
is the reappearance of corpuscular conceptions in a field from
which they had been excluded for nearly a century. The discovery
of the phenomena of interference and diffraction, the brilliant
theoretical work of Fresnel (already referred to) and the crucial
experiment—as it was thought to be—made by Fizeau and
Foucault when they measured the velocity of Light in water, had
seemed to furnish an irrefragable proof that Light consists of waves
in which energy is distributed continuously. As we have seen, the
old corpuscular conception of Light, which had rejoiced in the
eighteenth century in the support of Newton, had been abandoned
and almost forgotten at the end of the nineteenth century. To the
general amazement, however, it was destined to rise again from its
ashes: for the discovery was impending of certain phenomena
caused by Light and other rays hitherto unknown—phenomena
which could be explained only by a return to the corpuscular view.
The most important of these phenomena is the photo-electric effect,
which has been described previously.[2] Its laws differ completely
from those which the Wave Theory would have led us to expect,

[1] P. 61.   [2] P. 27.

and at first their interpretation seemed a matter of great difficulty; but as I have observed in my earlier discussion, Einstein saw that in order to explain the photo-electric effect recourse must be had, at least in some degree, to the corpuscular structure of rays. In 1905, accordingly, Einstein assumed that rays are formed of corpuscles transporting an energy varying inversely with the wave-length, and showed that the laws of the photo-electric effect followed readily from this hypothesis.

Other phenomena discovered still more recently—the Compton and the Raman effects—confirmed Einstein's hypothesis. It appeared that a certain number of facts were satisfactorily accounted for on the assumption that luminous energy is divided into the corpuscles called photons. This unexpected invasion of atomism in a sphere from which it was believed to have been completely banished caused the gravest difficulties to physicists. How was the new corpuscular concept of radiation to be reconciled with the voluminous and carefully investigated phenomena of interference and diffraction, for which the undulatory theory alone seemed capable of accounting? An examination of the reply to be made to this searching question was the source from which arose the strangely novel ideas of Wave Mechanics.

The only means of escape from the difficulties regarding Light was, in fact, to assume that it had two aspects: that it could be regarded as wave as well as corpuscle. Both aspects are revealed in the different classes of experiment, and the assumption was that they were two "complementary" aspects—Bohr's expression—of one and the same reality. When a ray exchanges energy with Matter, the exchange can be described as the absorption or the expulsion of a photon by Matter; but when, on the other hand, we wish to describe the motion of a group of light-corpuscles in space we must have recourse to the Wave Theory. If this idea is further developed we are led to the hypothesis that the density of a swarm of photons associated with a light-wave must of necessity be proportional at any given point to the intensity of this light-wave. Thus we are brought, if not to establish definitely, at any rate to entertain the notion of a kind of synthesis between the two

old rival theories of Light; a synthesis permitting of the simultaneous interpretation of interference and of the photo-electric
effect. The interesting aspect of this synthesis is that it shows that
in the case of Light, at any rate, waves and corpuscles are closely
linked to each other in the properties of the observed phenomena.

But if such is the case with Light, we may surely ask whether
the same does not also apply to Matter. A photon cannot be isolated
from its associated wave; and in the same way we may well assume
that corpuscles of Matter too are invariably accompanied by a
wave. And further, we ought surely to ask whether the somewhat
strange properties which the Quantum Theory had led us to attribute to the electron could not be interpreted on the assumption
that it has an undulatory aspect complementary to the corpuscular
aspect already familiar. Such were the weighty questions which
the reintroduction of corpuscles into the Theory of Light caused
physicists to ask; and from the examination of these questions
some ten years ago the fundamental concepts of Wave Mechanics
arose.

If, then, we boldly assume that waves and corpuscles are always
closely associated in Nature, then the motion of any corpuscle
must always be associated with the propagation of a wave. It must
be possible to express this connection in terms of the relations
between the two mechanical magnitudes characterizing the motion
of the corpuscle—energy and momentum—on the one hand, and
on the other the undulatory magnitudes—frequency and wave-
length—which are used to describe the propagation of the associated wave. Actually these relations can be established in a general
form within which the specific case of photons is subsumed; and
this general theory of the connection between corpuscles and their
associated waves is the foundation on which Wave Mechanics
has been built up. Naturally I cannot here describe in detail the
principles of this Mechanics: I shall confine myself to recalling
that it postulates that the length of the wave associated with a
corpuscle varies inversely with the velocity of the corpuscle: the
greater the velocity, the shorter the wave-length.

We can now comprehend some of the consequences of the new

Mechanics when applied to the electron. When the wave associated with a corpuscle moves freely in a region whose dimensions are great relatively to the wave-length, then the new Mechanics assigns to the corpuscle the same motion as would have been assumed by classical Mechanics. This is more particularly the case for the motion of electrons which we can observe directly; and that is why macroscopic investigation of electrons led observers to treat them simply as corpuscles. There are cases, however, where the classical laws of Mechanics fail to describe the facts. The first of these is when the motion of the associated wave is restricted to a region of space whose dimensions are of the same order of magnitude as those of the wave-length; and this is the case of the electrons within the atom. The associated wave is then forced to assume the form of a stationary wave; and Wave Mechanics shows that this stationary wave can only have certain wave-lengths which are strictly defined by the very conditions of the problem. According to the general principles of the new Mechanics, still further, certain possible energies for the intra-atomic electron correspond to these permissible wave-lengths of the associated wave. These states—the only ones possible, and possessing quite definite energy—correspond exactly with the quantized states of motion introduced by Bohr into his theory of the atom, and it is one first major success of Wave Mechanics that it explained the hitherto mysterious fact that these motions are the only ones which the electron within the atom can possess.

Another case where the motion of the electron is not presumed by Wave Mechanics to follow the classical laws of motion is when its associated wave meets with obstacles in the path of propagation. Under these conditions interference arises, and the motion of the corpuscle can no longer resemble that which it should have according to classical Mechanics. To form an idea of the consequences of this, we may follow the analogy of Light. Let us assume, then, that a ray of known wave-length is projected on an apparatus capable of setting up interference. We know that radiation consists of photons, and we can therefore also describe what we are doing by saying that we are directing a swarm of photons on the apparatus.

In the region where interference occurs, the photons are distributed in such a way that they are concentrated at the points where the intensity of the associated wave is greatest. If now we direct on the apparatus in question, not a ray but a stream of electrons having the same velocity, where the wave-length of the associated wave is the same as the wave-length of the ray used in the first instance, then we shall find that the wave interferes in the same way as in the first experiment, since it is the wave-length which determines the interference. Hence it is natural to assume that the electrons will concentrate at the points of greatest intensity of the associated wave—which is precisely the assumption made by Wave Mechanics. Hence in the second experiment the electrons ought to be spatially distributed in the same way as were the photons in the first—provided that the principles of the new Mechanics are correct. If then we can establish that this is in fact the case, we shall have demonstrated the existence of the wave associated with the electron and shall have provided the crucial experimental proof of Wave Mechanics.

This decisive experimental proof was actually obtained for the first time in 1927 by the two American physicists previously mentioned, Davisson and Germer, who sent a stream of electrons, having the same velocity, against a nickel crystal and thus produced phenomena exactly similar to those observed with X-rays. Subsequently the same phenomenon was observed and investigated by many other experimenters, notably by Professor G. P. Thomson; and today it is regularly obtained and utilized in laboratory work. The discovery of this striking phenomenon led to a complete and quantitative proof of the principles and formulae underlying Wave Mechanics.

We are thus in possession of direct proof of the view which holds that the electron is more than a mere corpuscle. It has its corpuscular, but simultaneously its wave aspect, and to understand the phenomena where it occurs it must be regarded as wave or as corpuscle according to the circumstances. Now how are these two aspects to be reconciled? I cannot here answer the question in detail: the reconciliation requires new and subtle concepts among which probability plays an essential part.

It must be added that it is not the electron alone which is both wave and corpuscle at the same time. This applies to the proton also, as more recent experiments have shown, and very probably to every material unit. Thus it is true of Matter as it is of Light that quite apart from the atomic, and discontinuous, aspect of the elementary entities there is the continuous and undulatory aspect, a discovery which has profoundly modified and greatly enriched the view we now hold of the electron.

* * *

Thus in the interpretation of the atomic world, Wave Mechanics has proved to be most successful, and its applications have completely changed the methods of microscopic Physics; its original form, nevertheless, suffered from two serious gaps. In the first place, it lacked Relativity Principles, and could therefore be applied only to electrons—or to other particles—of low velocity compared with that of Light; further, it did not assign to the electron the characteristic qualities of spin, while in its corpuscular aspect, again, it confined itself to treating it as an electrified material point. Clearly, then, it was necessary to find a more generalized form of Wave Mechanics, adequate to the requirements of Relativity Theory, and hence applicable to particles having every velocity without exception, and also embracing electron spin. It was Dirac who carried this development to a particularly happy conclusion. In its early form, Wave Mechanics represented the wave associated with the electron by a scalar function. Following a suggestion advanced by Pauli, however, Dirac assumes that the wave associated with the electron should be represented by a function of several components. His arguments led him to assume that the number of these components was four, and he succeeded in finding the four simultaneous partial differential equations which the four components ought to satisfy. While the original Wave Mechanics, therefore, represented the electron wave by a single scalar function satisfying an equation like the classical type of wave-equation, with partial derivatives of the second order, Dirac's Theory represents the associated electron wave by a function of four components, these four components satisfying a system of equations of the first order. The remarkable feature in this new Wave Mechanics of the electron is that once the equations governing propagation have been obtained, by very general arguments that do not concern spin, it is found that by this very fact spin, and all that follows from it, has been introduced. The essential point is that the study of Dirac's equations shows that they automatically involve the properties of angular momentum and magnetic moment, which had been assumed by Uhlenbeck and Goudsmit, being assigned to the electron. And while early Wave Mechanics, just like the older Quantum Theory, was unable to predict the anomalies of the Zeeman effect and the complexities of the fine structure of the spectrum, Dirac's equations enable these phenomena to be predicted with great precision. Thus Dirac's theory today provides us with the most complete idea of the electron we have. For on the one hand it attributes to it a corpuscular aspect, embracing mass, electric charge, magnetic moment and angular momentum; and on the other a wave aspect, which explains diffraction by crystals and the behaviour of the electron in atomic systems—an undulatory aspect in which the properties of spin manifest themselves by a certain anisotropy of the associated wave.

Attention should be drawn to one specific feature of Dirac's Theory which led its author to predict the existence of positive electrons—a prediction which was verified a little later. What I have in mind is the fact that Dirac's equations provide solutions involving negative energy, to which there would correspond electronic motion with paradoxical properties. Actually, the existence of this motion has never been proved. Here therefore there seemed to be a grave difficulty: the fact was that Dirac's Theory suffered from a plethora of potentialities. It was Dirac himself, however, who suggested a most ingenious escape from this dilemma. He had observed that according to Pauli's Exclusion Principle there can be only one electron, and no more, in any given state, and he assumed that in the case of electrons all possible states of negative energy are normally occupied in the Universe. Hence there followed a uniform density of electrons having negative energy, and Dirac assumed that this uniform density is quite incapable of being observed. To explain the existence of observable electrons, therefore, we should have to assume that there are in the Universe more electrons than are required to occupy all the states of negative energy, and that the surplus electrons occupy the states of positive energy, and thus form the totality of the electrons experimentally discoverable. Just at this point a new idea comes into operation which enables this concept—at first sight rather artificial—to lead to a great success. For there is nothing to prevent us assuming that one of the electrons, with negative energy, can pass into a state of positive energy under some external influence. In this case there would simultaneously appear an electron experimentally discoverable, and a hole or lacuna in the distribution of the electrons with negative energy. Now Dirac showed that such a lacuna behaves like a corpuscle having the mass of an electron, and an electric charge exactly equal to that of the electron but of opposite sign:—an anti-electron, or positive electron. Thus in certain exceptional cases it would be possible for a "pair" to be produced, consisting of one negative and one positive electron. No doubt Dirac's theory of "holes" would have left many physicists sceptical but for the fact that it received a striking experimental confirmation. The brilliant work of Anderson, and of Blackett and Occhialini, has actually shown that in certain exceptional circumstances (under the influence of cosmic rays) the existence of positive electrons is observed; and today these positive electrons or positrons are investigated in practically all laboratories. Dirac's ideas, still further, lead to the conclusion that positive electrons should be unstable and should tend to disappear on contact with Matter; and it is reasonable that if there should be a hole in the presence of a negative electron, the latter could fill the hole by a transition, at the same time emitting radiation; in which case the two electrons of opposite signs would have disappeared. The striking experiments by Jean Thibaud seem to have proved, and those of Joliot to have confirmed, this instability of the positrons.

Like the negative electron, too, the positive possesses a spin, and in all probability the case is the same for all elementary corpuscles of Matter—for example the neutron, which was discovered practically at the same time as the positron. It would seem probable, therefore, that elementary corpuscles universally possess not only the twofold aspect of corpuscle and wave, but also those properties of spin which no doubt are fundamentally bound up with the very existence of Matter. A deeper study of the electron's properties has thus led us to discover certain general characteristics which ought to belong to all elementary corpuscles, and of these characteristics Dirac's equations are at present the best mathematical expression.

* * *

We saw how the dualist view of Light, in which photons are associated with light-waves, serves as a guiding line in the structure of Wave Mechanics. The original aim of this Mechanics was to provide a general theory of the connection between waves and corpuscles—a theory applicable equally to Light and Matter, to photons and electrons. In its original form, nevertheless, Wave Mechanics is far from providing us with the foundation of an adequate theory of Light under its twofold aspect as wave and corpuscle. Why is this so? The first reason is that the original Wave Mechanics is not relativistic, and therefore is valid only for corpuscles of low velocity as compared with that of Light. Consequently it cannot be applied to the corpuscles of which Light itself consists. Secondly, the original Wave Mechanics employed a scalar and isotropic wave, and lacked the necessary symmetry elements required to explain the polarization of Light. Finally, it also fails to provide us with any means for giving to light-waves the electromagnetic character which, since the days of Maxwell and Hertz, we know that it certainly possesses.

With the introduction of Dirac's Electron Theory, however, the position has changed. For this is a relativistic Theory, and as such applicable to the photon. Further, it introduces an anisotropic wave, having a certain analogy with the polarization of Light. Finally, it connects electromagnetic magnitudes, derived from its intrinsic magnetic moment, with the corpuscle, and these magnitudes have a certain analogy with the fields of Maxwell's electromagnetic wave. It might thus have been hoped that an application of Dirac's equations to the photon would give us a satisfactory dualist theory which could be applied to Light. Actually, however, such was not the case, and without entering here into details I will merely say that a photon constructed on such lines would possess only half the symmetry necessary for an adequate theory of Light. Having made this discovery, the present author recently formulated a theory of Light in which the photon is regarded, not as a single Dirac corpuscle, but as a pair of Dirac corpuscles analogous to the pair formed by a positive and a negative electron. This conception leads to very satisfactory results, at any rate as far as the propagation of Light in empty space is concerned. It accounts also for the polarization of Light, and enables us to formulate exactly the real and deep relation subsisting between spin and polarization. We are also enabled to attach to the photon an electromagnetic field, completely identical with that by means of which Maxwell represented Light.

I do not, however, wish at this point to dwell on this new Theory of Light. More particularly, I will refrain from going into the question whether the two corpuscles, of which it assumes the photon to consist, ought not to be identified with the neutrinos, the existence of which is assumed by theoretical physicists in order to account for the apparent non-conservation of energy when continuous β-spectra are emitted by radioactive substances. I shall merely draw attention to the majestic curve which physical theory would have described, were this new theory to receive definite confirmation. For in that event, physicists would have begun from the simple idea of the electron regarded as a charged material point; they would then have been compelled, in order to explain quantum phenomena, to extend to the electron the dual nature discovered in Light, thus creating Wave Mechanics. In the next place, to include within Wave Mechanics the properties of spin which are necessary to explain a whole group of phenomena, they would have been compelled to complicate the new Mechanics by giving it the form due to Dirac. And finally, by a strange reversion to its first beginnings, the perfected Wave Mechanics would have returned to a point where it would serve in its turn in the formation of the dualist Theory of Light by uniting the photon, the light-wave, polarization and Maxwell's electromagnetic field in one harmonious whole.

But all such considerations are as yet hypothetical, and we must leave them. The present Chapter has shown how our conception of the electron developed, and in doing so became more complicated but also more rich—a process of 40 years. It would thus turn out, as is almost invariably the case, that we began by adopting too abstract and schematic a concept. It had therefore to be modified by the gradual introduction of new complexities, which frequently enough were disturbing to old modes of thought. But if we thus lost the fine simplicity of the beginnings, we have gained greatly in new knowledge and in the capacity to classify and to inter-connect physical facts on the atomic scale.

4

THE PRESENT STATE OF ELECTRO-MAGNETIC THEORY

(1) *General Survey of the Classical Electromagnetic Theory*

AT the end of the eighteenth century and in the first half of the nineteenth the work of Coulomb, Volta, Ampère, Oersted, Laplace and Faraday, to mention only a few among the most distinguished names, gave us a knowledge of the laws governing electric and magnetic fields, of currents and of the mutual relations between fields, charges and currents. Then came the great Clerk Maxwell, who generalized and systematized the laws formulated by his predecessors, and thus enunciated the electromagnetic Theory, the most striking characteristic of which is that it embraces the entire Theory of Light considered as an electromagnetic phenomenon.

The fundamental equations which are the foundation of the electromagnetic Theory are nothing else than the direct expression of the great experimental laws which Maxwell's brilliant intuition perfected by the introduction of the displacement current. I shall briefly summarize the form of these fundamental equations, using the Heaviside system of units, and denoting the universal constant equal to the velocity of Light in empty space by c. In the absence of any substance capable of electric or magnetic polarization these are:

$$\operatorname{div} \mathbf{H} = 0 \quad \operatorname{curl} \mathbf{H} = \frac{1}{c}\left(\frac{\partial \mathbf{E}}{\partial t} + \mathbf{J}\right)$$

$$\operatorname{div} \mathbf{E} = \rho \quad \operatorname{curl} \mathbf{E} = -\frac{1}{c}\frac{\partial \mathbf{H}}{\partial t}$$

where **E** denotes the electric field, **H** the magnetic field, **J** the "electric current density" vector and ρ the density of the electric charge defined on the macroscopic scale. The terms containing ρ and $\mathbf{J}$ in Maxwell's equations express the manner in which the presence of the charges and currents reacts on the electromagnetic field. In empty space ρ and $\mathbf{J}$ are zero, and it is easily proved by the help of the fundamental equations that electric and magnetic fields can in that case travel in transverse waves with velocity c. It was one of Maxwell's profoundest ideas to identify these waves with light-waves, and the discovery of Hertzian waves came several years later to confirm his views in brilliant fashion.

If it is desired to generalize the fundamental equations set out above, by extending them to substances capable of electric or magnetic polarization, we are led—a familiar fact—to introduce magnetic induction, $\mathbf{B}$, and electric displacement, $\mathbf{D}$. Very frequently these can be regarded as proportional at every point to the magnetic, and the electric, field respectively. In this case the fundamental equations are

$$\text{div } \mathbf{B} = 0 \quad \text{curl } \mathbf{H} = \frac{1}{c}\left(\frac{\partial \mathbf{D}}{\partial t} + \mathbf{J}\right)$$

$$\text{div } \mathbf{D} = \rho \quad \text{curl } \mathbf{E} = -\frac{1}{c}\frac{\partial \mathbf{B}}{\partial t}$$

These are also the analytic expression of the classical experimental laws.

The great laws of the electromagnetic Theory contained in the fundamental equations must further be completed by the laws expressing the mechanical action of the fields on the charges and currents. An electric charge e, at a point where the electric field is $\mathbf{E}$, is subject to a mechanical force equal to $e\mathbf{E}$; and an element of the current having intensity $\mathbf{I}$ and length ds, at a point where the magnetic field is $\mathbf{H}$, to a mechanical force equal to $(\mathbf{I} \times \mathbf{H})ds/c$.

In this way a system is obtained which represents completely the mutual relations subsisting between the fields, charges and currents as these are observed in normal large-scale experiment. The system thus formed satisfies the Principle of the Conservation of Energy and of momentum, provided that we attribute to the electromagnetic field an energy density equal to $(E^2 + H^2)/2$, and a momentum density equal to $(E \times H)/c$—at any rate outside polarizable substances. The hypothesis that momentum is localized in the electromagnetic field leads us to predict the existence of radiation pressure exercised by Light on any obstacles it meets. Such predictions have, as we know, been confirmed by certain extremely delicate experiments.

Maxwell's electromagnetic Theory, as extended and corrected by Hertz, met with great success in the field of large-scale phenomena, i.e. all those electromagnetic phenomena in which the microscopic structure of Matter has no place. The most famous of these successes was its prediction of Hertzian waves—the waves which extend the scale of light-rays and infra-red rays in the direction of increasing wave-lengths—and the interpretation of their properties. There is no need here to recall that the properties of the Hertzian waves have found a vast practical field in wireless telegraphy. In the form given it by Maxwell and Hertz, however, the electromagnetic Theory failed to explain fully the reactions between Matter and radiation—to describe, for example, the emission and absorption of radiation by material substances, or the way in which the presence of these substances affects the propagation of Light (diffraction, dispersion, etc.). In order to achieve such an explanation it was necessary to form some idea of the electrical structure of Matter on the microscopic scale, and of the way in which this structure reacts on electromagnetic radiation. Indications gathered experimentally, and perfected by the theoretical research associated with the famous name of H. A. Lorentz, have helped to fill this gap and to develop the electron Theory, which is an extension of Maxwell's and Hertz's electromagnetism.

In the electron Theory it is assumed that electricity has a corpuscular structure, and that all ponderable Matter contains a great number of extremely minute particles with an electric charge. We know further that experiments have demonstrated the existence of elementary particles of negative electricity, all of them resembling each other, the electrons. The electric charge and mass of electrons are expressed by extremely minute numbers, which however it has been possible to ascertain exactly. The electron theory regards material conducting substances as those in which the electrons are sufficiently free to travel *en masse* under the influence of an electric field, while non-conducting substances are those in which the electrons are confined to positions of equilibrium and can merely oscillate around these. The electric current is due to the motion of electrons in a conducting substance, etc.

This Theory was developed by Lorentz, who assumed that electrons have finite dimensions, and that a finite density of electricity ρ^* could be postulated internally. Hence he set out the general equations expressing the relations between electromagnetic fields and electrons in the form:

$$\operatorname{div} \mathbf{H} = 0 \quad \operatorname{curl} \mathbf{H} = \frac{1}{c}\left(\frac{\partial \mathbf{E}}{\partial t} + \rho^* \mathbf{v}\right)$$
$$\operatorname{div} \mathbf{E} = \rho^* \quad \operatorname{curl} \mathbf{E} = -\frac{1}{c}\frac{\partial \mathbf{H}}{\partial t}$$

where $\mathbf{v}$ is the velocity of electricity at the point where the microscopic density is ρ^*. It must here be noted that the quantity ρ^* is the microscopic density of electricity such as it must be defined when the existence of elementary charges is taken into account, while the density ρ used in Maxwell's equations is the mean microscopic density for any substance containing an immense number of elementary charges. To the above equations there must be added a supplementary one expressing the mechanical force exerted by an electromagnetic field on a unit charge having velocity $\mathbf{v}$. This force is

$$f = \mathbf{E} + \frac{1}{c}(\mathbf{v} \times \mathbf{H})$$

If a material substance containing a great number of elementary charges of the same sign (electrically charged, that is, in the usual sense of the term) or a material substance containing electrons all in a state of motion (or through which an electric current is passing, in the usual sense of the term) is considered, then naturally Lorentz's equations lead to the Maxwell-Hertz macroscopic equations, which are in the same form. Clearly this agreement was necessary in order that Lorentz's theory should prove admissible; yet the fact must not be overlooked that in reality Lorentz was making an extremely bold extrapolation when he adopted, for the microscopic and individual equations of his electron Theory, a form analogous to that of Maxwell's macroscopic and statistical equations. Actually Maxwell's equations were suggested directly by experiments on electrified substances and currents, i.e. on phenomena where an immense number of elementary particles are involved. Lorentz's formulae then assert that laws of the same form are applicable to the relations subsisting between fields and electric charges, when the microscopic distribution and the corpuscular structure of electricity are considered in detail: and, as I have just said, this is a bold extrapolation.

Nothing is to be gained by expatiating on the successes achieved by the electron Theory. What most struck the scientific public some 40 years ago was no doubt its prediction of the Zeeman effect; yet it must not be forgotten that Lorentz's Theory has provided an excellent interpretation of the dispersion of Light, has explained the emission of radiation by Matter by means of the 'acceleration-wave' given out by electrons when their velocity varies, has elucidated the mechanism of the conduction of heat and of electricity in metals by the assumption that they contain free electrons, etc.

The development of Einstein's Theory of Relativity did no harm at all to the electron Theory: the latter can easily find room for the implications of the Principle of Relativity. More particularly the variation of the mass of the electron with its velocity, which is implied in the dynamics of Relativity, was found by Guye to harmonize well with experiment. By adapting itself to the relativistic form, indeed, the electron Theory actually found that it automatically escaped from certain difficulties encountered in the interpretation of such experiments as the famous negative attempt made by Michelson to demonstrate the absolute motion of the Earth in Space by a local optical phenomenon.

(2) *Difficulties Encountered by the Electromagnetic Theory*

The electromagnetic Theory and the Theory of electrons which—
by extending its application to the microscopic scale—complements
it, rendered the greatest services to physicists and furnished the
explanation for a very large number of phenomena. Finally,
however, they came to the end of their applicability and encountered
great difficulties.

The germs of a first and grave difficulty are contained in the
very concept of the electron. One might conceive of the electron
as a mathematical point surrounded by an electric potential having
the form $\frac{e}{r}$; but the actual form of the equations placed by Lorentz
at the foundation of his Theory assumes that it is possible to speak
of the electric density inside electrons, which implies that they must
be imagined as corpuscles which, though admittedly of extremely
minute dimensions, are nevertheless finite. Again, if the electron
were of the nature of a geometrical point, the energy of the electrostatic field surrounding it would be infinite, which appears to be
physically impossible. Accordingly the electron has been envisaged
as a sphere having a finite radius r_0, within which the electricity
is distributed in a certain way. Assuming then that the entire mass
of the electron is of electromagnetic origin, and taking the values
experimentally discovered for the charge and mass of the electron,
we are led to attribute to it a radius $r_0 = 10^{-13}$ cm. But in this
case the great difficulty is to understand how such a sphere, containing electricity all of the same sign, can exist in a stable manner,
since its constitutent parts ought to repel each other. We are
driven to imagine, with Henri Poincaré, that there is a pressure at
the surface of the electron coming from outside and preventing
the particle exploding; but we remain wholly at a loss to explain
the origin of such a pressure.

A good many more difficulties, connected more or less directly
with the existence of quanta, were encountered by Maxwell's and
Lorentz's theory of electromagnetism. Thus the interpretation of
the emission of electromagnetic waves by Matter, according to

which this phenomenon is due to the accelerations of the electrified corpuscles which it contains, accounts perfectly for the phenomena when a huge number of corpuscles is concerned, as is the case when a Hertzian wave is sent out by a wireless antenna; the absorption by Matter of these electromagnetic waves is also easily explained. But since Planck's Theory of black body radiation came to be developed, and still more since Bohr's Atomic Theory, we know that the mechanism by which elementary corpuscles emit radiation while being accelerated as described by the classical Theory cannot be correct. Atoms are systems formed of electrons revolving around a central nucleus, and simultaneously intensely accelerated; and they are capable of possessing quantized stationary states in which they emit no radiation whatever; and all this is wholly in contradiction with the laws of the electron Theory. It is, in fact, only when an atom suddenly changes its stationary state, and passes from energy E_n to energy E_m, that it emits radiant energy in the form of a quantum of radiation of frequency ν and of energy $h\nu = E_n - E_m$. (Bohr's Law of Frequencies.) Naturally the classical Theory, as enunciated by Maxwell and Lorentz, is wholly unable to find a place both for the existence of stationary states with no radiation, and also for the sudden transitions from one stationary state to another accompanied by the emission of quanta.

The old ideas also fail to give any account of the important laws which the investigation of spectra has shown to exist. The prototype of these laws was that which states the frequency of the lines in Balmer's Series. All these laws satisfy one general principle, viz. Ritz's Combination Principle, according to which there is for every kind of atom or molecule a series of spectral terms of such a kind that all the frequencies of the lines, emitted by the atom or molecule in question, are differences between two of these spectral terms. The electromagnetic Theory, on the other hand, if applied to the motion of electrons on the atomic scale, would always lead to a continuous emission of energy, a process accompanied by a progressive diminution of the corpuscular motion; the lines emitted would be diffuse, and the only relations possible between their frequencies would be harmonic and with no regularity analogous to Ritz's Principle. Here then there was a complete contradiction of the old theories of the electron; and the only phenomenon connected with the emission of Light by Matter which is satisfactorily explained by the electron Theory is the normal Zeeman effect. But here it must be observed that the Zeeman effect only exceptionally presents the so-called normal aspect: most often it is abnormal, in which case it manifests peculiarities which the electron Theory cannot interpret. The Quantum Theory, again, had no difficulty in arriving at the formulae for the normal Zeeman effect, for here the quantum treatment leads to the same result as the classical. The abnormal effect, on the other hand, long remained inexplicable even by the Quantum Theory, and it is only the assumption that the electron has a magnetic moment (Uhlenbeck's and Goudsmit's spin hypothesis) that has enabled us to understand its real origin.

I have pointed out the difficulties encountered by the electron Theory of electromagnetism in predicting the reactions between Matter and Radiation phenomena on the microscopic scale. These difficulties prove that the terms depending on ρ^* and $\rho^* v$, which occur in the second terms of Lorentz's equations, do not adequately account for the creation of electromagnetic fields by elementary electric charges and their motion, and this despite the fact that the terms depending on ρ and $\mathbf{J}$ in the second terms of Maxwell's equations account for the creation of electromagnetic fields by charged bodies and currents on the macroscopic scale with great exactitude. One might have hoped, therefore, that at least the equations for electromagnetism in empty space would have been correct. These equations are as follows, both in Lorentz's and in Maxwell's system:

$$\operatorname{div} \mathbf{E} = 0 \quad \operatorname{curl} \mathbf{H} = \frac{1}{c} \frac{\partial \mathbf{E}}{\partial t}$$

$$\operatorname{div} \mathbf{H} = 0 \quad \operatorname{curl} \mathbf{E} = -\frac{1}{c} \frac{\partial \mathbf{H}}{\partial t}$$

and, as we know, they do actually provide an excellent explanation of the undulatory propagation of Light *in vacuo* and of other types of radiation, regarded as electromagnetic disturbances. Unfortunately, the fact that the emission and the absorption by Matter of radiation having the frequency ν always takes place in quanta of energy equal to hν — as has been shown in the course of the development of the Quantum Theory and its ample experimental confirmation—suggests the idea that radiation must have a corpuscular structure even in empty space. The discovery of the photo-electric effect and its interpretation by Einstein, his further theoretical research on the fluctuations of energy in black-body radiation, the more recent discovery of the Compton effect and various other developments, have progressively inclined physicists to revert, to a certain extent, to the old views held by Newton and to assume that radiation has a corpuscular structure. Maxwell's equations for empty space, however, appear unable to account for this corpuscular structure (i.e. for the existence of light-corpuscles or photons); here again the electromagnetic Theory suffered shipwreck.

(3) *The Correspondence Principle and Wave Mechanics*

Under the form which Lorentz had given it, then, the electromagnetic Theory was unable to explain how elementary material systems emit and absorb radiation. It is quite unable, also, to interpret the existence of the stationary states of atoms, and the sudden emission of radiation in the form of a quantum of energy when they pass from one stationary state to another. Hence it would at first seem that the classical Theory can do nothing whatever to help us in investigating this type of phenomenon. Some 20 years ago, however, Bohr fortunately perceived that despite its inability to give an exact description of the facts, the classical Theory of electromagnetic radiation could still prove of some use. For he realized that the classical Theory was valid on the macroscopic scale for phenomena involving a great number of electrons, and that for this reason it ought to give the exact statistical results of the emission of quanta of radiation by electrons whose quantized motion corresponds to very high quantum numbers; as regards the motion of electrons corresponding to medium or low quantum numbers, however, we can no longer expect that the classical Theory will give us absolutely exact information, although one can still hope that it may give us certain qualitative indications. Such are the contentions which, stated in more precise form, constitute Bohr's Correspondence Principle. In the older versions of the Quantum Theory, this Correspondence Principle had remained somewhat qualitative; it has, however, proved extremely useful in predicting, not only the emission and absorption of radiation by atoms, but also the influence exerted by Matter on its propagation—e.g. in such phenomena as those of diffraction and dispersion.

Somewhat later still, the Correspondence Principle was given its exact form, thanks to the development of the new Quantum and Wave Mechanics. In its Wave form, in the first place, the new Mechanics consists essentially in the principle that to predict the motion of material corpuscles, it is necessary to consider the propagation of the waves associated with these corpuscles. Further, in the more abstract Quantum form given it by Heisenberg and Dirac, the new Mechanics consists in the substitution of special numbers—matrices or q numbers—for the quantities characteristic of the old Mechanics in the various calculations. These do not always fulfil the Commutative Rule in multiplication, but they do satisfy the same formal equations as the corresponding quantities of the older Mechanics. From the calculation of q numbers it is possible to deduce—by methods which I will not set out here— the values of the mechanical magnitudes observable experimentally. Different as they may appear at first sight, however, the two forms of the new Mechanics are found to coincide in the last analysis. The profound significance of the new Mechanics became apparent gradually, thanks particularly to a close study of the concept of measurement and of the possibilities of experimental determination made by Bohr and Heisenberg. In this way it was seen that the non-commutation of the two q numbers, representing two mechanical magnitudes, corresponds to the fact that it is impossible to measure these two magnitudes simultaneously with perfect exactness, a fact which can also be easily expressed in the Wave formulation of the new Mechanics. In this way the co-ordinate x of a corpuscle and the corresponding component p_x of the momentum—two magnitudes which correspond to q numbers which do not commute—can never be simultaneously measured exactly. Still further, it can be shown that the uncertainty Δx in the measurement of x, and the uncertainty Δp_x in that of p_x, are always such that we have

$$\Delta x \cdot \Delta p_x \geqslant h$$

where h is Planck's constant, whose finite value appears here as limiting the possible exactness of simultaneous measurements. This is Heisenberg's famous Principle of Uncertainty; its analogue will be found at a later stage in Quantum electromagnetism.

We are now in a position to explain how the new Mechanics has allowed us to state the Correspondence Principle in exact terms; and in my explanation I shall make use of the terminology of Wave Mechanics. The wave to be associated with a corpuscle is represented by a function, Ψ, which is treated—at any rate for a first approximation—as a complex scalar magnitude. We are then led to assume that the intensity of the wave, equal to the product of Ψ and the conjugate complex quantity $\overline{\Psi}$, gives us, for any given instant at any given point, the probability that the corpuscle is at that point at that instant. If the corpuscle is an electron with charge e, then the quantity $e\Psi\overline{\Psi}$ will represent a kind of probable mean density of electricity in the region where the electron is moving, and the variations of this quantity will enable us to define a kind of mean electric current. The statement of the Correspondence Principle can be formulated exactly by making use of this mean density and mean current to calculate the energy radiated by the motion of the electron, the equations of the classical Theory governing the radiation of energy being utilized for this purpose. Let us consider an electron in an atom. According to Wave Mechanics, there are *stationary* waves Ψ, corresponding to the stable quantized states of these electrons. Thus the stable state of energy E_j corresponds to a stationary wave Ψ_j, so that

$$\Psi_j = a_j(x, y, z)e^{\frac{2\pi i}{h}E_j t}$$

The state of the atom will always be capable of being represented by a function Ψ, such that the latter is the sum of the functions Ψ_j, since the totality of the functions Ψ_j constitute what mathematicians call a complete system.

Accordingly for an atom, assumed for the sake of simplicity to contain only a single electron, the function Ψ can be written

$$\Psi = \sum_j c_j \Psi_j$$

For the mean density of the negative electricity in this atom we shall then have the value:

$$e\,\Psi\overline{\Psi} = e \sum_j \sum_k c_j \overline{c_k}\, a_j \overline{a_k}\, e^{\frac{2\pi i}{h}(E_j - E_k)t}$$

to which there corresponds an electric moment $\mathbf{M}$, the component of which parallel to one of the axes (the axis of x for example) will be

$$\mathbf{M}_x = e \iiint \Psi\overline{\Psi}\, x\, dx\, dy\, dz = e \sum_j \sum_k c_j \overline{c_k} X_{jk}\, e^{\frac{2\pi i}{h}(E_j - E_k)t}$$

with

$$X_{jk} = \iiint a_j(x, y, z)\, \overline{a_k}(x, y, z)\, x\, dx\, dy\, dz$$

The quantity X_{jk} is what Heisenberg calls the element having indices j, k, of the matrix corresponding to the co-ordinate x. If we assume that the radiation of the atom can be deduced from the variations of $\mathbf{M}$ by the classical formulae giving the radiation from a distribution of electricity having a variable total electric moment, then it appears that the atom ought to emit frequencies

$$\nu_{jk} = \frac{E_j - E_k}{h}$$

where the component of frequency ν_{jk} vibrates parallel to the axis of x with an intensity proportional to $|X_{jk}|^2$. We thus encounter once again Bohr's Frequency Law; it agrees with the Combination Principle, and at the same time we obtain an exact rule which enables us to predict polarization and intensity. Such is the way—a rather intuitive one—in which Schrödinger has suggested that Wave Mechanics ought to state the Correspondence Principle. But this intuitive method has the drawback that it might suggest inexact ideas; and indeed the way in which we have defined the electric moment M of the atom might suggest that it is one and the same atom which radiates all the frequencies v_{jk}. Such is not the case, however, and it would completely contradict the existence of stable quantized states, and also Bohr's fundamental idea, which is that radiation is associated with the transitions from one stable state to another. Actually, the sum in the second formula giving M_x merely represents the totality of the possibilities of emission, together with their respective probabilities. The square of the matrix element X_{jk}, therefore, represents the relative value of the probability that an atom, which at first is in a stable state of energy E_j, will pass into a stable state of energy E_k, with the emission of a quantum hv_{jk} polarized parallel to the axis of x. Heisenberg's Quantum Mechanics postulates this meaning of X_{jk} directly, without using the intermediate link of Schrödinger's mean electric density, which might be a source of error.

Schrödinger himself, in fact, insisted on the difficulties involved in taking literally his representation of the mean electric density by the expression $e\Psi\overline{\Psi}$. One of these difficulties in particular is the fact that though this mean density can be used to calculate the radiation emitted by an atom, it cannot in any way be used when we have to calculate the influence of an external electromagnetic field upon the atom. For this influence cannot be predicted in the least by calculating the influence of the external field in question upon the mean distribution of electricity given by the density $e\Psi\overline{\Psi}$. To obtain precise results we must introduce the electromagnetic field into the equation for the propagation of the wave associated with the electron; and this can be expressed approximately by saying that the external field affects by no means the hypothetical distribution of electricity of density $e\Psi\overline{\Psi}$, but the electron itself.

Most of the difficulties enumerated here can be removed if we use the so-called method of "second quantization" (*Überquantelung*), suggested by Dirac, Klein, Jordan and other authors. The essence of this method is to regard the wave-functions Ψ as being themselves q numbers and as not obeying the Commutative Rule. In a fairly recent Paper, Heisenberg has shown that it is possible in this way to satisfy the requirements of the Correspondence Principle without becoming involved in contradiction. I shall not insist here, however, on this method of second quantization: its results are interesting, but of an extremely formal character.

(4) *The Quantum Theory of the Field*

We have seen that in the course of its development the new Mechanics enabled us to state in more exact terms the Correspondence Principle, and to continue to use the general laws of the radiation of electromagnetic energy by charges in motion in the form in which they follow from the Maxwell-Lorentz equations. Yet though these equations can be used in practice even on the microscopic scale by virtue of the Correspondence Principle, we must nevertheless face the fact that actually the entire structure of electromagnetic Theory today needs to be rebuilt. We saw that Maxwell's equations are not correct even in empty space, because they do not account for the existence of photons. An extremely interesting attempt to formulate a new electromagnetic Theory, compatible with the concept of the photon, has been suggested by Heisenberg and Pauli under the name of The Quantum Theory of the Field. Since it was first formulated a good deal of work has been done on it, among which I may mention especially that of Rosenfeld and Solomon.

I have pointed out that the deepest reason why, in the calculations of Quantum Mechanics, mechanical magnitudes must be represented by q numbers not generally obeying the Commutative Rule, is that here mechanical magnitudes cannot generally be regarded as capable of simultaneously being measured exactly. In the Quantum Theory of the Field, too, electromagnetic magnitudes are treated as q numbers; and here also this assumption is necessitated by the fact that the magnitudes of the field cannot generally be taken as simultaneously being measurable exactly in Quantum electromagnetic Theory. Actually, Heisenberg has shown that the very existence of photons in an electromagnetic field compels us to introduce the Principle of Uncertainty in the case of electric and magnetic fields, and this, just as in the new Mechanics, limits the exactness of the simultaneous measurement of a co-ordinate and of the corresponding component of momentum. Heisenberg supported his view by an analysis of the methods used in the measurement of such fields. To explain more fully this new Uncertainty Principle, let us denote by E_x one of the components of the electric field, and by H_y a component of the magnetic field perpendicular to E_x. In this case Heisenberg has shown that if the fields are measured in a region having the dimensions ϵ, then the uncertainty ΔE_x in E_x, and ΔH_y in H_y, must always satisfy the inequality:

$$|\Delta E_x \cdot \Delta H_y| \geqslant \frac{hc}{\epsilon^4}$$

The new electromagnetic Theory was constructed by Heisenberg and Pauli in analogy with Quantum Mechanics. Their method was to extend the formal methods of the new Mechanics to the case of fields defined at every point in space. The mathematical formulation of the Quantum Theory of the Field is somewhat elaborate, and I shall not set it out here in detail. Suffice it to say that we find equations, for these fields, having the classical form of Maxwell's equations, with the difference that the magnitudes of the field are q numbers.

So long as we are dealing with a vacuum, i.e. with an electromagnetic field in empty space, the new Theory seems satisfactory enough. It accounts most adequately for the existence of photons, and indeed extends the concept of the photon by introducing the idea of the "longitudinal photon" corresponding to the electrostatic field in addition to the "transverse photon" corresponding to the classical electromagnetic waves. But even in this region of pure electromagnetic fields a difficulty was encountered at the very beginning. For the electromagnetic field should have a minimum energy corresponding, in the theory of black body radiation, to the existence of a residual energy at absolute zero— a circumstance of which Planck already had a presentiment. The difficulty was that this residual minimum energy should apparently have an infinite value. It proved possible, however, to turn this obstacle, for Rosenfeld and Solomon succeeded in causing the infinite energy at absolute zero to disappear by an ingenious modification of the Theory.

The difficulties with which the new electromagnetic Theory met when dealing with the reactions between fields, charges and currents seem to be much graver. I have previously observed that classical Theory was compelled to assign to the electron an infinite self-energy if it wished to regard it as a geometric point, and that the absurdity of this conclusion was easily avoided by taking the electron to be a sphere having a finite radius. On the other hand, the Quantum Theory is wholly unable to define the radius of the electron, and for this reason it seems that it must again encounter the problem of the electron with infinite energy. This is not the only unwelcome result involved by the Quantum Theory of the Field in its present form, and one of the physicists who have done most to amend it, Rosenfeld, has shown that it leads to absurdities in the case of the spectral terms of atoms and the gravitational energy of the photon. Faced by these reverses, he admits that from the formal point of view the new Theory has proved misleading.

(5) Conclusion

To sum up, it is certain today that despite its repeated successes in the macroscopic domain, and despite its usefulness on the microscopic scale due to the application of the Correspondence Principle, the electromagnetic Theory is not fundamentally correct and must be reconstructed. Unfortunately, the only important attempt to find a substitute[1]—the Quantum Theory of the Field—appears to have turned out a failure. And recently it has been realized that this collapse is only one aspect of the difficulties which stand in the way of an attempt to import the methods of the new Mechanics into the sphere of Relativity, i.e. into those regions where the velocity of Light, c, cannot be considered as extremely great. These difficulties are entirely unexpected, since it was the Principle of Relativity which led to the laying of the very foundations of Wave Mechanics; but they are incontestable. More particularly, the best form of the new Mechanics in which Relativity is admitted is certainly Dirac's brilliant theory of the magnetic electron. It has led to many satisfactory results; yet it is certain that despite the confirmation derived from the discovery of the positive electron, this theory is still liable to objections which it has failed to remove. Today it would seem that the Quantum Theory of electromagnetism is of necessity faced by the same type of fundamental difficulty, since it is its aim to transplant the methods of Quantum Mechanics into a domain where it seems absolutely impossible not to take Relativity into consideration.

Recent research, all of which seems to have its origin in Bohr's complex and profound theories, tends towards the demarcation of a limit for the validity of the methods of the new Mechanics, and even for the employment of such fundamental notions as those of Time and Space; but even research of this kind, whose nature at present is somewhat negative, fails to tell us in what direction and by what methods the genuine new electromagnetic Theory, in which there is room for the existence of Quantum phenomena, must develop.

[1] Since these lines were written (in 1932) other attempts have been made, notable among them being the fine Non-linear Theory of Electromagnetism due to Born. Nevertheless the real solution of the problem still remains to be found.

III

LIGHT AND RADIATION

I

A SURVEY OF THE HISTORY OF OPTICS [1]

THE history of modern optics begins in the seventeenth century.

Of course this does not mean that the scientists of the seventeenth century had no forerunners in the sphere of optics. In the evolution of the human mind it is wrong to make too clean a cut, and there can be no doubt that no step was ever taken in the advance of Science which was not based on the work of earlier periods. Setting aside, however, certain questions of priority which could be resolved only by specialist studies, one may fairly say that the real advancement of modern optics dates from the first half of the seventeenth century and may largely be identified with the great name of René Descartes.

Some time before Descartes, Snell had experimentally discovered the ratio of the angles in refraction; but it is to Descartes that the credit belongs for having been the first to state in exact terms the system of laws of reflection and refraction, to which, at any rate in France, we attach his name. To interpret these laws Descartes adopted the corpuscular conception of Light, and assumed that these corpuscles meet with a resistance which deflects their trajectory, whenever they enter Matter which is dense in the optical sense.

Rather earlier in the seventeenth century Pierre de Fermat, a magistrate and a geometrician, had shown that Descartes' laws could be deduced from a Principle of Minimum Time,[2] to which he attributed a teleological significance. According to this well-known Principle, the path of a ray of light passing through two given points A and B is always such that the time taken by the ray in passing from A to B is less than that which would

---

[1] Extracts from the inaugural lecture of a course on The Interaction between Matter and radiation. (November 1933.)

[2] Or, more accurately, Stationary Time.

correspond to any other path, however closely similar to that of the ray.

About the same time experiments met with new successes. While Hooke and Grimaldi observed the colours of thin metallic films without making any attempt at interpreting them, Newton discovered in 1666 the spectral decomposition of Light by the prism, and the Danish astronomer Roemer, observing the occultations of the satellites of Jupiter, inferred in 1676 that the velocity of Light in empty space must have a finite value. Pursuing a different method, Bartholin discovered double refraction in spar crystals. All these experimental discoveries provided a powerful impulse to theoretical speculation. Christian Huyghens (as I have already remarked) first of all advanced the Wave Theory in a definite form by assuming that there is an ether in which light-waves can move. With the help of the Principle which today bears his name, he showed that the Wave Theory can explain the phenomena of reflection and refraction, and also succeeded in finding an explanation for double refraction. But, admirable as were his labours, they did not command universal assent. They suffered, moreover, from a serious lack: they offered no explanation for the fundamental fact that Light travels in a straight line. Newton for his part adopted the corpuscular view, and showed the advantages it possessed for a dynamic interpretation of the propagation of Light in a straight line, of reflection and of refraction. At the same time he was already acquainted with certain interference phenomena (Newton's rings) and accordingly, by a bold stroke of intuition, tried to establish an association between waves and corpuscles—the motion of a projectile and the propagation of a periodicity. But this theory of "Fits of easy reflection and easy transmission" was too far ahead of his time; it remained embryonic and underwent no further development.

The majority of eighteenth-century scientists followed Newton and adopted the corpuscular theory. At the same time there were some notable exceptions, among whom Euler may be mentioned.

We have seen that at the beginning of the nineteenth century there came some fresh experimental discoveries to add an impetus to the more sedate pace of the development of optics. In 1801 Young made some accurate observations of interference phenomena, and enunciated their ruling principle; but his work was rather of an empirical nature, and at first little notice was taken of it. About the same time Malus observed certain polarization phenomena, but without entirely clearing up their obscurities. In this manner, however, the way had been prepared for a new advance in theory, an advance due to the genius of Augustin Fresnel (1788–1827), who rejected the corpuscular Theory which at this stage was still being defended with talent and authority by Laplace and Biot. As I have previously observed, Fresnel reverted to Huyghens' Wave Theory, which he completed by adding an interpretation of the motion of Light in a straight line, and gave an account of interference and diffraction phenomena of which he also made an extensive experimental investigation. The opposition school, led by Poisson, was worsted when demonstration came to prove the correctness of even the most paradoxical predictions of the Wave Theory. Some time later Fresnel adopted a suggestion made by Young and introduced the principle of the transversality and the polarization of light-waves, being in this way enabled to develop his brilliant theory of the intensity of reflection and of refraction. The period following Fresnel's death was characterized by the gradual triumph of his ideas. Fizeau's and Foucault's experiments (1850), which showed by direct measurement that the velocity of Light in water is less than in empty space, seemed to provide a crucial proof of the Wave Theory's correctness.

But though triumphant in the experimental sphere, this Theory met the greatest difficulties in the theoretical domain when it sought to become a full and complete mechanical theory of ether vibrations. Despite the efforts made by many able theorists—Poisson, Green, MacCullagh, F. Neumann and, later, Lord Kelvin, Carl Neumann, Lord Rayleigh and Kirchhoff—it remained impossible to establish definitely a coherent theory of ether vibrations. It was under a wholly different form, and one implying a much more

far-reaching surrender of any kind of visual representation, that
the Wave Theory was developed about the year 1870, about which
time Clerk Maxwell created the electromagnetic Theory, a more
exact form of ideas originally due to Faraday. Maxwell's Theory
is based entirely on the rather abstract concept of the electromagnetic field; and all attempts to reduce this concept to that of a
certain state of a hypothetical medium (the electromagnetic ether)
have had to be finally abandoned. Maxwell showed that Light can
be embraced within the general category of electromagnetic disturbances—a brilliant conception by which he was enabled to
subsume the whole of optics under the electromagnetic Theory.
From this new point of view—a more formal one than that of
Fresnel—the fact that Light is of the nature of a wave finds expression in the fact that the electromagnetic fields of the light-
wave are certain periodic functions of the co-ordinates of Time and
Space.

Any really complete theory of optical phenomena, however,
assumes a knowledge of the laws of interaction between Light and
Matter, since it is impossible to study Light otherwise than through
the effects it produces on Matter. A comprehensive theory of this
kind had been the goal pursued by the champions of an elastic
ether; but it did not become a reality until H. A. Lorentz developed
his electron Theory. By introducing the concept of the electron,
and the Law of the interaction between the electromagnetic field
and electrons, Lorentz and his rivals were enabled to investigate
the different ways in which Matter reacts when an electromagnetic
light-wave passes through it. The electron Theory, by stating in
more precise terms the meaning, and by substantially extending
the scope, of the fragmentary results reached by earlier theories,
succeeded in finding a place for the formulae of dispersion, the
laws of the absorption of Light by absorbing bodies, those of the
various magneto-optic and electro-optic effects, etc. Its success in
interpreting, too, the normal Zeeman effect is familiar. But fruitful
as was this attempt to analyse the interactions between Matter and
radiation, it encountered totally unforeseen difficulties when it
came to the details of these phenomena. On entering the atomic

domain, the electromagnetic Theory, together with Lorentz's complementary electron Theory, encountered the quanta: the result was that the very nature of Light once more became matter of debate.

* * *

The electromagnetic Theory, still further, and the electron theory of the interactions between Matter and radiation, lead to incorrect results regarding thermal equilibrium. For they yield a law of spectral distribution of the energy of radiation in thermal equilibrium (Rayleigh's Law) which contradicts the experimental results for high frequencies and which, in a sense, is absurd, since it leads to an infinite value for the total density of energy. These perplexing consequences of the classical Theory were avoided by Planck through the introduction of the entirely novel idea that the only way in which Matter can emit radiant energy is by quanta equal to $h\nu$, where ν is the frequency emitted and h a new universal constant. It followed from Planck's hypothesis that Matter could lose energy only in finite quantities. This did not necessarily mean that a ray, once given out, must have a discontinuous structure, for it was possible to develop the theory in two different ways as far as the absorption of radiation by Matter was concerned. The first point of view—a more straightforward one, which eventually proved the more acceptable—consisted in assuming that the elementary constituents of Matter can only assume certain quantized states of energy, whence it follows that in absorption, as well as in emission, the exchange of energy between Matter and radiation occurs in quanta.

But this necessarily implied that radiation has a discontinuous structure. Shrinking from this disturbing consequence of his own ideas, however, Planck persisted in making every effort to work out a different and less radical form of Quantum Theory, in which emission alone would be discontinuous, while absorption would remain continuous. Matter would then be capable of accumulating continuously part of the radiant energy falling on it, but it would be able to emit it only discontinuously and in finite quantities.

The aim of Planck's efforts is readily seen: he wished to preserve the continuity of radiation, because this principle alone seemed compatible with the Wave Theory which was supported by so many successes. But despite all the ingenuity brought by Planck to the development of this second version of the Quantum Theory, the latter was disproved by the eventual advance of our knowledge.

An essential stage in this advance was the discovery of the photo-electric effect, and its interpretation by Einstein. The facts have been outlined already; and Einstein showed that they can be explained only by a certain reversion to the corpuscular view of Light, in which any ray of frequency ν is considered as being formed of corpuscles containing energy equivalent to $h\nu$. This is the theory of light-quanta or photons. Supported first of all by investigation of the photo-electric effect, and later by various ideas developed by Lorentz and Einstein—the statistical equilibrium between the molecules of a gas and the surrounding radiation in thermal equilibrium, fluctuations of energy in black body radiation —this view demonstrated the need for accepting the Quantum Theory in its first and more radical form.

Actually it was the Quantum Theory in its first version which prevailed with the triumph of Bohr's Atomic Theory.—According to Bohr an atom possesses certain "stationary states" in which it emits no radiation. When it passes from one stationary state to another (the quantum transition), its energy suddenly varies by a finite quantity. If the transition is accompanied by a diminution of energy, the atom emits a finite quantity of energy in the form of a quantum $h\nu$, such that the frequency of the ray emitted is equal to the quotient of the diminution of the atom's energy, when the transition occurs, by Planck's constant, h. This is Bohr's Frequency Law. But under the influence of a ray of frequency ν the atom can also undergo the reverse change, and pass from one stationary state to another such state having a higher energy— the result of absorbing a quantum $h\nu$, provided that the difference between the energies of the respective stationary states is exactly equal to $h\nu$.

Such a view of the emission and absorption of radiation by

Matter is obviously in complete contradiction with the electro-magnetic Theory evolved by Maxwell and Lorentz. According to the latter a planetary atom, as envisaged by Rutherford and Bohr, ought to be emitting radiation continually, its frequency being continuously variable. Like classical Mechanics, classical electromagnetism also is proved to be incorrect in the microscopic sphere, though there should be a certain compromise on passing from the microscopic to the macroscopic—as is the case with Mechanics. Such was the idea clearly formulated by Bohr in 1916—a few years after the development of his atomic Theory—in the form of his "Correspondence Principle." Bohr begins by pointing out that there are very many quantum numbers for which the difference between the energies of the consecutive stationary states tends to become infinitely small, so that there is a tendency for continuity to reappear. The frequencies predicted by the Quantum Theory in such cases accordingly tend to agree with those of the classical Theory. He accordingly *assumes* that the intensities of emission and of polarization, as calculated by the classical Theory, should remain correct also for the Quantum Theory in this sphere; then, by a bold extrapolation, he assumes that even for the smaller quantum numbers the indications fur-nished by the classical Theory may be used to a certain extent. Bohr's Correspondence Principle proved extremely useful some fifteen years ago in enabling intensities and polarizations to be predicted within the sphere of the Quantum Theory as it was then formulated. More particularly, it enabled selection rules to be stated, indicating which among the spectral lines predicted by Bohr's Frequency Law have an intensity above zero, and therefore are really capable of observation.

During recent years, as I have repeatedly observed, there has been a development of Wave Mechanics. We now regard radiation from two points of view—as corpuscular and also as undulatory; and this view has helped in the formulation of the new Mechanics, whose essential idea is to assume that Matter also has the same double wave-corpuscle character. Optics has thus served as a guide in the erection of the structure of Wave Mechanics. A very curious

thing, however, has happened: Wave Mechanics—at least in its
non-relativistic form—very soon reached a high degree of perfection, while the dual Theory of Light—the photon Theory and
the Fresnel wave Theory—lagged far behind. The chief reason for
this is that any theory of Light must necessarily be relativistic,
because the corrections which Relativity introduces in the dynamics
of a corpuscle increase in importance as the velocity of the corpuscle
approximates to that of Light. The theories advanced during recent
years in order to quantize the electromagnetic field, so as to find
in it a place for photons (Dirac's Theory and the Quantum Theory
of the Field enunciated by Heisenberg and Pauli), are not wholly
satisfactory and seem to be rather provisional. To grapple with
the problems connected with the interactions between Matter and
Light, physicists were usually content to employ an extension of
the Correspondence Principle; and indeed the new Mechanics
lends itself to a more precise statement of this Principle than those
sanctioned by the older Quantum Theory. To reach this end a
combination is effected between the ideas of classical electromagnetism and the magnitudes of Wave Mechanics. It is certainly
a somewhat awkward method: it does not account for the dual
character of radiation, and it leaves the concept of the photon
rather obscure. At the same time it has proved possible in this
way to obtain interesting results and to form a theoretical schema
of a certain number of the phenomena connected with the interaction between Matter and Light—a schema which, if probably
provisional, is undoubtedly of practical use.

2

OLD WAYS AND NEW PERSPECTIVES IN THE THEORY OF LIGHT

THE story of the different theories of Light is one of the most exciting parts of the history of Physics. It is here, more clearly than perhaps anywhere else, that the century-old struggle between the corpuscular and the undulatory conceptions of Light has shown how contradictory hypotheses, both based on experimental facts, can both contain a part of the truth, and how the advance of Science has on many occasions been effected through the synthesis of opposed points of view. Nor has the long story yet come to an end; for at the present moment the Theory of Light, which at one time played so large a part in inspiring the modern dualist theories of Matter, is lagging behind these theories, and the future no doubt holds further shocks and further developments in store for it.

* * *

The corpuscular Theory of Light was based on certain simple facts familiar practically throughout history: the propagation of Light in a straight line, and reflection. The interpretation given by this Theory is direct and appeals to the eye, and many of the greatest scientists, headed by Newton himself, have supported it during the course of centuries.

Nevertheless, since Newton's time the corpuscular Theory has always been faced by a severe obstacle in the shape of the phenomena of interference and diffraction. Newton himself had discovered the phenomenon of the coloured rings which are named after him, and to interpret it had developed his theory of "Fits of easy reflection and easy transmission" which, regarded for long as a complicated and hybrid conception, would today be considered rather as the anticipation by a man of genius of current theories. Actually, the experimental research on diffraction and interference at the beginning of last century by Fresnel and Young had to be carried out, and the theoretical labours to which we have already found the name of the former to be attached had to be completed, before the Wave Theory triumphed over its rival—a state of affairs which was destined to endure for a long period. Fresnel (to repeat) reverted to and perfected the arguments previously outlined as having been used by Christian Huyghens, the seventeenth-century champion of the Wave Theory; and as has already been observed, he succeeded in showing that this theory explains the propagation of Light in a straight line, reflection, refraction in its various forms, diffraction and the interference phenomena. It is true that for the first two the Wave Theory gives a less simple and direct account than does the corpuscular; but when we come to refraction the former has the better of it, and for interference and diffraction it is the only theory to provide an adequate interpretation. But Fresnel—a brilliant physicist— gave us additional knowledge of the utmost value about Light in the phenomena of polarization, and thus revealed an aspect of fundamental symmetry appertaining to light-waves. For a light-wave must be defined, not only by a scalar "light variable," but also by an oscillating vector, which in the simple case of a plane monochromatic wave in a vacuum, or in an isotropic medium, is in the wave-front normal to the direction of propagation. On the basis of the polarization of Light Fresnel constructed the theory of the intensity of reflection at the surface of separation of two substances and also that of the propagation of Light in anisotropic media. Experiments have brilliantly confirmed the theory, and it is to be found with hardly a change in every modern work.

In elaborating his Wave Theory Fresnel—to repeat my earlier observations—had made use of the idea of an elastic ether of such a kind as to be able to transmit only transverse vibrations. In this way it was possible to obtain results completely agreeing with experiment over a wide field. But there were also difficulties and contradictions; still further, there was no apparent connection between luminous phenomena and those of electromagnetism, in which research was at the time advancing rapidly. Faraday's discovery in 1850 of the first magneto-optic phenomenon, however—the rotation of the plane of polarization of Light passing through a substance in which a uniform magnetic field prevails—drew attention to one necessary connection between these two groups of phenomena. Next came Clerk Maxwell, who formulated that admirable synthesis, the Electromagnetic Theory of Light. Here the luminous vibration is represented by the oscillating electric vector propagated in waves. This vector is accompanied by a second, the magnetic vector, which also oscillates, and which in the simple case of a plane monochromatic wave in empty space is equal and perpendicular to the first, both being in the wave-front normal to the direction of propagation. The velocity of propagation in a vacuum is equal to the constant c in Maxwell's equations, i.e. to the ratio of the electromagnetic unit of charge to the electrostatic unit. All this is familiar and agrees with experiment. In this way Maxwell has given to Light an electromagnetic structure which it is impossible to neglect.

Unhappily, the fine synthesis effected by Maxwell was challenged by the sudden discovery of phenomena where a structure of Light differing completely from that of the electromagnetic wave seemed to manifest itself. The first of these was the photo-electric effect. Its essential characteristic is that Light of frequency v seems able to transfer its energy to Matter only in finite quantities proportional to the frequency. We recall that by a brilliant intuition Einstein realized (1905 and onwards) that this experimental fact meant the necessity of reverting, to some extent, to the corpuscular Theory of Light. He assumed that Light of frequency v is divided into corpuscles of energy $W = hv$; and he was thus led immediately to the experimental law of the photo-electric effect, which establishes a connection between the kinetic energy of the electrons photo-electrically expelled, and the frequency of the incident radiation.

What does this new corpuscular Theory, considered apart from any idea of a wave, give us? It obviously yields an immediate and quasi-visual interpretation of the rectilinear propagation of Light and of reflection, and it can be developed a little further if we apply Relativity dynamics to the photon—which is clearly necessary for a corpuscle moving with velocity c. Actually, if we apply Relativity dynamics to the photon we obtain a satisfactory explanation of radiation pressure and of the various kinds of Döppler effect. Finally, the new corpuscular Theory won a brilliant success on the day when the Compton effect was discovered. This consists of a lowering of frequency by the scattering of X-rays by Matter, a change in frequency which is easily interpreted by the assumption that the incident photon is scattered by its encounter with an electron originally at rest. By this encounter energy and momentum are transferred from the photon to the electron, with the result that the energy and hence the frequency of the photon is lowered, provided that we assume that $W = h\nu$. Thus the theory, based on the idea of the photon, accounts quantitatively for the observed effects.

Yet despite the success of the theory of photons in the interpretation of a considerable number of phenomena, a radically corpuscular theory of radiation cannot be maintained, because it cannot account for the numerous and exactly observed experiments where we find diffraction and interference phenomena. The entire body of research carried out by Fresnel and his followers prevents us from returning to a pure "theory of emission."

*   *   *

Such was the deadlock reached by the Theory of Light when Wave Mechanics arose; and its fundamental idea can be represented in a whimsical form as follows: the Theory of Light was suffering from a strange disease, which manifested itself in the rival dualistic forms of Fresnel's and Maxwell's waves on the one hand and of photons on the other! To alleviate the situation, then, the desperate remedy might be tried of inoculating the Theory of

Matter, immune thus far, with the same disease! Actually, however, there was a substantial reason for such a course; for the fact was that the Theory of Matter too had been manifesting alarming symptoms for a number of years. The fact that it had proved necessary to quantize the motion of particles of Matter, which had first become evident in the theory of black body radiation— a method which had achieved a striking success in Bohr's theory of the atom—showed that the existence of the quantum of action prevented us from extending the concepts and equations of classical Mechanics to the atomic field. The presence of integers in quantum formulae endowed the latter with a certain similarity to the formulae of interference in the Wave Theory; and the analogy between the Principle of Least Action—the keystone of classical Mechanics—and Fermat's Principle—the keystone of geometrical optics—suggested that classical Mechanics might be an approximate form of a more general Wave Mechanics, standing with regard to the latter in the same relation as that in which geometrical optics stands to wave optics. In this way arose the notion of extending to Matter the dualism of corpuscles and waves which had proved necessary in the case of Light. In Wave Mechanics, accordingly, it was assumed that a complete description of the material units could not be effected by operating with corpuscles alone, and that this idea had to be supplemented by that of waves, the correspondence between the dynamic description of the corpuscle in uniform motion and the frequency and wave-length of the associated wave being based on the fundamental relations which include the theory of photons as one particular case. In this way a satisfactory synthesis became apparent, in which the corpuscular and the wave aspect—two aspects interconnected by the same general relations—must be taken into account simultaneously, both for Light and for Matter. So satisfactory is this synthesis, and so completely does it form the foundation of Wave Mechanics, that it must be maintained at all costs. We shall, however, shortly see the difficulties met with as soon as we try to express in exact terms the affiliation between the photon and the ultimate constituents of Matter; but I am none the less convinced that it would be an aberration to attempt the erection of a fresh barrier between the Theory of Matter and that of Light.

I must here add the fact that the fundamental idea of Wave Mechanics received complete experimental confirmation by the discovery of the diffraction by crystals, first of electrons, and later of protons and heavy nuclei.

*   *   *

In revealing the undulatory aspect of the constituents of matter, particularly of electrons, Wave Mechanics undoubtedly brought the two Theories—of Matter and of Light—closer together. The question still remains, however, whether it brought about a complete synthesis—a single Theory applicable to both Matter and radiation. In its original form, certainly, it did not. In the first place, this form rests on non-relativistic equations. Primitive Wave Mechanics, in other words, is a development of Newtonian Mechanics, not of Einsteinian. It is indubitably applicable only to material particles whose velocity is much lower than c, and hence, with equal certainty, it cannot be applied to the photon. This alone shows that if we wish to find a single Theory to embrace both Matter and Light, we must generalize the original Wave Mechanics in the relativistic sense.

Another difference between the original Wave Mechanics and the Theory of Light is the fact that the former contains no element of symmetry corresponding to polarization. Here again one feels that, to reach a single Theory of Matter and Light, some quantity corresponding to polarization must be introduced, which is absent from the original form of Wave Mechanics.

Finally, I must insist on the unique part played by the photo-electric effect in the Theory of Light. Material corpuscles can exchange energy and momentum by interaction; but though they can thus lose their entire kinetic energy, they invariably conserve their mass-energy, and never disappear.[1] In the photo-electric effect, on the other hand, the photon loses its entire energy on coming into contact with Matter, and is annihilated. The interchange of

[1] To this, however, the recently discovered dematerialization of positive and negative electrons forms an important exception.

energy and momentum between the photon and Matter in such phenomena as the Compton effect, or reflection by a rotating mirror, is quite similar to the exchanges in collisions between material corpuscles; but the photo-electric effect is wholly different, and the difference cannot be over-emphasized. Now it must be noted that whenever an interference or diffraction phenomenon is observed, the observation is effected through the medium of a photo-electric effect, which takes place in the sensitive layers of the retina in the case of direct observation, or in a film of gelatine in the case of photographic observation. If we wish to interpret these phenomena in terms of the interferences occurring in an electromagnetic field connected with the photon, we must establish a close connection between the transition between states of the photon corresponding to the photo-electric effect—a transition which may be compared to an annihilation of the photon—and the electromagnetic field.

Another obstacle must finally be mentioned which seems to stand in the way of an amalgamation of the theories of Matter and Light. This obstacle consists in the fact that the elementary corpuscles of Matter—electrons, for example—follow Fermi-Dirac statistics when they form a large group, whereas photons, when they form such a group, as in black body radiation, follow Bose-Einstein statistics. The mention of this difficulty, to which I shall revert later, must here suffice.

Unable thus to embrace Matter and Light within one single Theory, what has Wave Mechanics done to deal with the problems of the interaction between Matter and radiation? It has had recourse to a hybrid method, relying on the Correspondence Principle, where the corpuscular nature of Light does not explicitly appear. The method has certainly yielded very interesting results and provides a good approximate prediction of the phenomena under investigation; but to my own mind it has one grave fault, that of completely obscuring the symmetry in the structure of Light and Matter respectively. There can be no doubt that Dirac's photon Theory comes nearer the truth; but this Theory can deal only with large groups of photons by the method of second quantization, whereas I think it probable that it should be feasible to deal with individual phenomena without making use of this method.

*   *
*

First of all then, and to achieve any real advance in the photon Theory, a relativistic form of the new Mechanics was needed to provide for the corpuscle an element of symmetry of the same kind as polarization. Now this new form had in fact been in existence for a number of years, in the shape of Dirac's Theory of the magnetic electron. Personally, I experienced great satisfaction at the appearance of this Theory of the magnetic electron, for I had always felt that it would contribute greatly towards the formation of a single Theory of Matter and Light. Actually, it offers us on the one hand a Wave Mechanics of the corpuscle which is as relativistic as Quantum Theory permits, given the present state of Science. Thus the first difficulty mentioned above is greatly diminished. Still further, Dirac's Theory automatically introduces a quantity, the spin of the electron, which at first sight has a certain affinity with polarization, whence we can at least hope that we may reduce the difference between material particles and photons resulting from the polarization of Light. For the electron, and more generally for every particle obeying his equations, Dirac assumes the properties of magnetic moment and angular momentum. These intrinsic properties of the electron are in accord with the idea of the magnetic rotating electron suggested at an earlier date by Uhlenbeck and Goudsmit, and they enable us to explain the anomalies of the Zeeman effect and of the fine structure of optical and of Röntgen spectra. It is altogether satisfactory, therefore, to find that these essential properties are automatically contained in Dirac's equations.

I cannot here develop Dirac's Theory, which requires a complicated mathematical apparatus, nor enumerate the many successes with which it has met. I must, however, insist on one peculiarity which was thought at first to constitute a difficulty for his Theory, but which appears since to have turned out to be a support. I allude to the fact (previously mentioned) that Dirac's equations yield solutions corresponding, for the electron, to states of motion having negative energy.[1] Now these states of motion would have properties of a completely paradoxical character (for example, by checking an electron with negative energy we should increase its velocity); and it is certain that motion of this kind has never in fact been found to exist. Thus (to repeat) we are apparently faced by a grave difficulty: in one way, in fact, Dirac's Theory seems to suffer from an *embarras de richesse*. But I have just referred to my earlier discussion of the ingenious method of removing the obstacle, suggested by Dirac himself, and based on Pauli's Exclusion Principle. From these considerations, therefore, one general idea arises: for any corpuscle obeying Dirac's equations, whatever may be the values of charge and mass, there must be a corresponding anti-corpuscle, which stands to the corpuscle in the same relation as does the positive electron to the negative electron.

*   *   *

Revenons à nos photons. To obtain a single Theory of Light and Matter, the simplest notion would be to compare the photon to a corpuscle obeying the Dirac equations, but having an electric charge and mass both equivalent to zero, or at any rate quite negligible in comparison with even the mass and charge of the electron. In such a case, the photon would have an appreciable energy that permitted its existence to be manifested experimentally, only when it possessed a velocity equal to, or at any rate indistinguishable from *c*.

There is, however, a serious *a priori* objection to any identification of the photon with a Dirac corpuscle. The position is that it is assumed today, for quite good reasons, that an elementary corpuscle, having the properties of a Dirac electron, obeys Fermi statistics in the same way as any other complex particle formed of an odd number of such corpuscles. On the other hand, the complex particle formed of an even number of Dirac corpuscles obeys

[1] cf. p. 98.

Bose-Einstein statistics. Now the photons observe these statistics; and it is quite certain that they do so, since otherwise we could not have Planck's Law for black body radiation. Photons must therefore consist of an even number of elementary corpuscles.

To this primary objection, encountered by an attempt to compare the photon with a Dirac corpuscle, others are added when we try to develop his Theory a little further. I cannot here, however, discuss details, but I may add that a photon constructed in this way would, in a manner, have only half the symmetry needed in order to allow us to associate with it an electromagnetic field of the Maxwellian type. A Dirac corpuscle, in other terms, can only provide us with half a photon.

The idea thus suggests itself that the photon might be considered as consisting of two Dirac corpuscles. But we know that Dirac's Theory, completed by the idea of lacunae already dealt with, makes a positive anti-electron correspond to the negative electron. More generally, we can make an anti-corpuscle correspond to every corpuscle obeying the Dirac equations, the former being defined as a hole or a lacuna within a domain of negative energy. On such a view it becomes tempting to imagine the photon to consist of a corpuscle having a negligible mass and charge and obeying the Dirac equations, associated with an anti-corpuscle of the same kind. It is a hypothesis to which we have been recently led, and it is an attractive one. For it is reasonable to suppose that a photon constituted in this way should be capable of annihilation in the presence of Matter, transferring to it at the same time the whole of its energy—the annihilation corresponding to a quantum transition by which the corpuscle contained in the photon fills up the accompanying lacuna. Actually such a transition, accompanied by annihilation, would constitute the photo-electric effect, whose fundamental importance from the theoretical point of view has already been stressed, while the electromagnetic field associated with the photon would then have to be defined as a function of this transition. Actually it is possible to show that an electromagnetic field, completely analogous to that which in Maxwell's system defines the luminous wave, can be associated with this transition leading to annihilation. In itself this is an encouraging fact; and further, since the photon is now assumed to consist of a corpuscle and an anti-corpuscle both of which are defined by the Dirac equations, the photon ought to follow Bose-Einstein statistics, which experiments show that it actually does follow.

To construct a photon after the schema outlined above we must assume the existence of a class of corpuscles obeying the Dirac equations and having either no electric charge and mass, or at any rate a charge and mass negligible as compared with those of the electron, minute as the latter are. Now there are in fact certain indications supporting the existence of this new physical entity. When β-rays are emitted by the nucleus of a radioactive substance, the Principle of the Conservation of Energy is not, apparently, satisfied. We may therefore well sacrifice this important Principle of Conservation so far as nuclear phenomena are concerned; and this is the solution supported by Bohr's great authority. Alternatively, we may assume that the phenomenon of the emission of β-rays by radioactive nuclei is accompanied by the emission of a new kind of particle, which it would be hard to detect experimentally because of the slightness of its action on Matter. The energy carried by these particles would thus escape experimental detection, at any rate with such means as we possess today, and on this hypothesis we could retain the Conservation of Energy. This idea was advanced some time ago by Pauli and Fermi, who called the new—and hypothetical—type of corpuscle the neutrino. Certain recent research has rendered the existence of the neutrino more probable, although there is not yet any apparent means of establishing it by direct observation. Francis Perrin and Fermi have shown that if the neutrino does exist, its mass must be zero, or at least negligible compared with that of the electron. At the same time it would be impossible to identify the neutrino with the photon, since it has so far escaped experimental detection, so that its action on Matter must be extremely slight. In other words, it can have no electromagnetic field. This naturally suggests an identification of the neutrino with that corpuscle having no mass which forms part of the photon, and the neutrino would thus be a kind of semi-photon. In a state of isolation, i.e. when not accompanied by an anti-neutrino, it would have no electromagnetic field, since it could not be annihilated by the photo-electric effect; but when united with an anti-neutrino it would form a photon and would have an electromagnetic field of the Maxwellian type.

It must be admitted that these ideas are still largely hypothetical and raise numbers of difficult questions, and before they can be accepted they must undergo a thorough scrutiny. Yet it does seem probable to me that if some day a satisfactory theory of the nature of the photon is constructed, it will show elements of similarity with the outline I have just drawn.

*   *   *

The prolonged antagonism between the Corpuscular and the Wave Theories as applied to Light, deriving fresh stimulus from the discovery of the photo-electric effect, had seemed, with the development of Wave Mechanics, to end in a comprehensive synthesis. For Light, as well as for Matter, it seemed that waves and corpuscles were two complementary aspects of a single physical reality. It looked as though a single theory, comprehending both Matter and radiation, and amalgamating waves and corpuscles, were within reach. Actually, however, this comprehensive theory has not yet been formulated. In its application to Matter, Wave Mechanics developed to the accompaniment of striking successes; but it lacked the relativistic aspect and the symmetry necessary to enable it to include Light. Dirac's Theory of the magnetic electron was a most important advance, because it introduced into Wave Mechanics relativistic invariance on the one hand, and on the other those elements of symmetry which recall the polarization of Light. There can be small doubt, therefore, that when a theory to embrace both Matter and radiation is formulated, it will be based on this Theory of Dirac's. Recent discoveries, still further, have introduced us to those new physical entities, positive electrons and neutrons, to which I have had as yet no occasion to refer, and have led us to suspect the existence of others like the neutrino. No doubt it is on such lines that we shall gather the data needed to understand the character of the photon. The Theory of Light, then, has a long and striking history; and a fine career lies before it.

3

AN INSTANCE OF SUCCESSIVE SYNTHESES
IN PHYSICS:
THE VARIOUS THEORIES OF LIGHT

To succeed in grasping the manner in which the human mind
proceeds when dealing with the multiplicity of facts with which it
is faced in any attempt to co-ordinate and interpret the phenomena
of Nature, it is instructive to select some special class of phenomena,
and to trace through the history of Science the way in which the
successive views adopted by men to picture and to classify them
have been transformed. The result of such a study is to reveal one
of the capital difficulties with which scientific theories are faced.
It is this. The same physical entity, though in its innermost nature
it quite certainly is one and one only, may reveal itself to us under
such varying aspects that we are compelled, in order to describe
them, to form successively different and sometimes contradictory
views about the entity in question. The labours of successive
generations of scientists acquaint us with a growing number of
phenomena, and we are enabled, to the same degree, to group them
in separate classes, each of which requires that a full description
of the entity under investigation shall embrace a certain characteristic quality. At times the position becomes very serious—for
example when it is found that the different characteristics of the
entity, revealed successively in the different groups of phenomena,
appear to be incapable of being reconciled within the framework
of any single theory. Yet the postulate which lies at the root of
every scientific inquiry, the act of faith which has always sustained
scientists in their unwearying search for explanation, consists in
the assertion that it must be possible—though perhaps at the heavy
cost of surrendering ideas held for long and concepts of proved

usefulness—to reach a synthetic view uniting all the partial theories suggested by the various groups of phenomena, and embracing them all despite their apparent contradictions. In this way we see clearly before our eyes, with its difficulties, its passing reverses and also its splendid triumphs, the goal of the endeavour of theoretical Science; an effort directed towards synthesis and union, which strives to reduce to a kind of intellectual singleness the immense complexity of the facts.

The history of the different theories of Light is an admirable instance of the circumstances I have just mentioned. What dominates this history is the secular controversy between the corpuscular and the Wave theories of Light. As we have found in earlier Chapters, there are certain phenomena which strongly suggest that Light should be treated as consisting of corpuscles in rapid motion, i.e. as having a discontinuous structure; while others suggest that it should be compared to a wave moving in an elastic medium, where energy spreads continuously on the surface of a continuous wave. It is a curious fact that a number of phenomena appear to hesitate, if it can be put that way, to declare themselves for either side; they can be equally well interpreted by either of the opposed views and remain neutrals in the battle. We know that at one time it seemed as though this battle had gone to the Wave Theory; but the discovery of the quantal and discontinuous structure of Light put everything in doubt again, and it became necessary to look for a comprehensive theory to reconcile waves and corpuscles. Here Wave Mechanics came to the rescue, though eventually the synthesis which it effected was less complete, from the point of view of the Theory of Light, than might have been at first expected. The reason is that there are certain groups of phenomena, apart from those in which corpuscles, and those in which waves are revealed, and apart also from those which appear as neutrals in the strife between these two opposed conceptions, which give to Light certain peculiar characteristics. We have in mind polarization phenomena and electro-optic phenomena, i.e. those phenomena which show that Light has an electromagnetic character. It is a familiar fact that a place has been found by suitable adjustments for these qualities of Light within the Wave Theory. Polarization was brought within the Wave Theory by Fresnel, who assumed that light-vibrations are transverse and not longitudinal, as had originally been assumed in accordance with the views of Huyghens. In other words, Fresnel represented Light by a vector in the plane of the wave, instead of representing it by a scalar quantity—a "light-variable"—as had previously been done. As regards the electromagnetic nature of light-waves, this had been foreseen by the genius of Maxwell, who elaborated an electromagnetic interpretation of the Wave Theory of Light; and indisputable facts have come to support the correctness of his intuition. In its original form Wave Mechanics contains nothing analogous to the polarization of Light, nor does it offer us any chance of connecting a corpuscle to an electromagnetic field analogous to Maxwell's luminous electromagnetic field. It is for this reason that Wave Mechanics could not produce a completely comprehensive theory of Light. It could certainly explain how the corpuscular and the wave interpretations could coexist as explanations of luminous phenomena; but it leads to a theory analogous to that of Fresnel's "light-variable" and provides no explanation of polarization nor of the electromagnetic character of Light. But the brilliant Theory of the magnetic electron formulated by Dirac, which completes Wave Mechanics by assigning to the elementary corpuscles new mechanical and electromagnetic properties in the shape of spin and of magnetic moment, has supplied the new Mechanics with exactly those elements which were lacking, and has enabled it to form a complete Theory of Light in which there is a place for both aspects of Light discovered at different times. It is true that this comprehensive Theory has not yet been finished in all respects; yet there is reason to hope that today the end is in sight. This point will be dealt with at the conclusion of the present Chapter.

*   *   *

Let us revert to a detailed examination of the successive theories of Light. Here we might follow the chronological order in which these theories appeared; but I consider it more interesting to show how optical phenomena fall into five leading groups, to which the various successive forms of the theories adopted by physicists to interpret the qualities of Light correspond.

The first class of luminous phenomena which men observed is formed—to judge from the standpoint of modern knowledge—of what I have called "neutral phenomena," i.e. those which are equally well interpreted by either the corpuscular concept or the undulatory. There is first the motion of Light in a straight line, which manifests itself by the existence of luminous rays and by reflection in mirrors—simple phenomena which are obviously and immediately explained by comparing Light to corpuscles obeying the principle of inertia and capable of reflection like elastic bodies when obstacles are encountered. On the other hand, their wave interpretation, though less obvious, is also completely satisfactory, as is proved by Huyghens' explanation of reflection, and by the famous—and decisive—arguments produced by Fresnel concerning rectilinear propagation. Thus the undulatory and the corpuscular theories rank as equals in the interpretation of these two simple and fundamental phenomena: both are acceptable, though for simplicity the corpuscular view is easily superior. There is next the phenomenon of refraction. Qualitatively this too has been known for a long time: quantitatively it has found an interpretation thanks to the research of Snell and Descartes. This phenomenon too is neutral; it is equally well interpreted by corpuscular and by wave ideas—for example, once again, on the lines followed by Huyghens. Yet there is here an important difference between the two theories. The undulatory defines the index of refraction, as occurring in Descartes' laws, as the ratio of the velocity of Light in empty space to its velocity in the given refracting medium, while the corpuscular hypothesis defines the index of refraction by the inverse ratio. For long the exact velocities were unknown, so that either definition could be accepted. Towards the end of the nineteenth century, however, the advance made by experimental technique allowed Fizeau and Fresnel to measure the velocity of Light in different media, and their experiments proved

that the velocity in refracting media having an index higher than
1 (glass, water, etc.) is lower than in empty space, which seemed to
amount to a crucial proof in favour of the Wave Theory. A new
discussion of these results, however, made today in accord with
Wave Mechanics, shows that it furnishes arguments only against
a corpuscular hypothesis based on Newtonian Mechanics, but
not against a corpuscular theory based on the conceptions underlying the various new types of Mechanics. Incidentally, this shows
how greatly the alleged finality of certain experiments depends on
the validity of the theories used to interpret them. Ultimately, then,
refraction must be classified among the neutral phenomena, though
the wave interpretation maintains a certain advantage. Other and
less easily observable phenomena, whose discovery is of more
recent date, may be placed in the same category, among them
radiation pressure and the various Doppler effects. Provided that
Relativity dynamics is applied to light-corpuscles, a natural procedure today, the corpuscular view gives a very good interpretation
of the facts. But it is well known that the Wave Theory explains
them equally well, so that they must be classified as neutral. Finally
the spectral distribution of energy in the case of radiation in thermal
equilibrium, as expressed by Planck's famous Law, falls within the
same category. On the one hand, Planck's Law can be obtained
in such a way as to represent the distribution of the energy of
light-corpuscles in a gas in a state of thermal equilibrium, provided
that we apply a suitable statistical method (Bose-Einstein method);
on the other, we can arrive at the same law in such a way as to
make it the expression of the distribution of quantized luminous
energy between the various monochromatic stationary waves
which may be set up in the region under consideration. In this case
Jeans' classical arguments are employed. At the same time there is
always here a more or less openly avowed combination of the
ideas of waves and of corpuscles—a combination which cannot
really be justified unless the synthetic notions of Wave Mechanics
are introduced.

We thus have an extensive category of luminous phenomena
which can be interpreted equally well by the idea of corpuscles or

by that of waves. Let us pass to a second category of luminous phenomena—to those which may be considered as specifically undulatory. Chief among these are the phenomena of interference and of diffraction, those discovered by Newton (Newton's rings) having been longest known; their discoverer was led, in order to interpret them, to add to his corpuscular view of Light certain periodic elements, and thus by his "theory of fits of easy reflection and easy transmission" produced a sort of premonition of the synthesis which was actually realized two centuries later by Wave Mechanics. Essentially however, as has previously been observed, it was Young and Fresnel who, at the beginning of the nineteenth century, discovered the vast category of interference and diffraction phenomena. Augustin Fresnel took up, and with the sweep of his peculiar genius completed, the ideas which Huyghens had put forward 150 years earlier. In its first form, Fresnel accounts for this entire group of phenomena by his theory of light-waves, in which Light is likened to an elastic longitudinal wave and is represented by a scalar magnitude—the "light-variable." Fresnel's scalar Wave Theory interprets the phenomena of interference and diffraction, in which waves play a specific part; and apart from these novelties (as they were in his day) it interprets the entire group of neutral phenomena—rectilinear propagation, reflection and diffraction—which had long been known. Today all this is familiar and is to be found in every treatise on optics.

Soon however Fresnel had to go further in order to embrace a third large class of facts, namely those in which the polarization of Light occurs. The polarization of Light had been accidentally discovered by Bartholin in 1660 and by Huyghens in 1690 during their experiments on Iceland spar. It was definitely established by Malus in 1810, since when it was carefully studied by contemporary physicists, particularly by Fresnel himself. To these phenomena in which polarization appears we shall give the name of "vectorial" luminous phenomena. All of them show that Light has not the same qualities at every azimuth around the direction in which it is moving; and any given beam of Light can be broken up by suitable apparatus into beams having different symmetries.

In the case of a beam which is plane polarized, the Light ought to be characterized in a vacuum by a vector vibrating in a fixed direction perpendicular to the direction of motion, i.e. in other words by a direction situated in the plane of the wave. In a more complete form of his theory Fresnel introduced this vector concept into the undulatory Theory, attributing to the light-wave the character of a transverse wave and thus completely explaining all the phenomena of polarization. One of the more striking inferences from this was his brilliant theory of the propagation of Light in anisotropic media, which is to be found almost unchanged in all present-day works. Among the most significant characteristics of Fresnel's theory is the assumption that all light-vibrations must be transverse, since otherwise there would be a transverse and a longitudinal light having very different properties, which in fact is not the case. Mathematically this transverse—and inevitably transverse—quality of Light is explained by the fact that the divergence of the light-vector is zero. In the mechanical theory of Light, within which Fresnel and those who came after him have tried to assimilate light-vibrations to the vibrations of a hypothetical elastic medium, the ether, which is supposed to penetrate all substances, this zero divergence implies that the medium has an infinite rigidity. This conclusion, however, is not easily reconciled with the fact that the heavenly bodies are entirely unhampered in their movement through the ether, and it proved to be one of the great difficulties in the way of this mechanical interpretation of the vectorial theory of light-waves. This mechanical interpretation, in fact, has long been abandoned, and I shall mention it no more. I shall observe, however, that Fresnel's vectorial undulatory theory was an excellent synthesis, since it gave an explanation of the neutral phenomena, of the specifically undulatory phenomena and of the polarization phenomena of a vector character—i.e. of practically the entire body of facts relating to Light known a century ago.[1]

Let us pass to a fourth class of luminous phenomena, the discovery of which goes back to the middle of last century. In 1846

[1] cf. pp. 125, 132.

Faraday discovered the effect which bears his name, viz. the rotation of the plane of polarization of Light when it moves in a refractive medium under the influence of a magnetic field. At the time this discovery was of capital importance, because it showed for the first time that there was a connecting link between electro-magnetic and optic phenomena—thus uniting two spheres which appeared to be alien to each other. A few years later Clerk Maxwell, guided by Faraday's work, developed his famous electromagnetic Theory. In his endeavour to sum up analytically the known body of facts relating to electricity and magnetism, he formulated the equations which express in mathematical terms the Laws of induc-tion and of the creation of electric fields by charges and of magnetic fields by currents, and lastly the non-existence of "true" magnetism. After having drafted this system of equations, however, he was himself the first to realize that it was unsatisfactory, because it led to difficulties in connection with the conservation of electricity and the difference between open and closed circuits. And now, by a brilliant piece of intuition, he completed his system by adding certain terms which, as he thought, represented a displacement current existing even in empty space, and capable of abolishing the difference in question between open and closed circuits. Having in this way obtained a satisfactory system of electromagnetic equa-tions, he found that these implied that electromagnetic disturbances must travel in a vacuum in the form of transverse waves at a constant velocity equal to the ratio of the units of charge in the electromagnetic and electrostatic systems. This ratio is usually represented by the letter c. At this point there occurred to him the brilliant idea that Light itself is an electromagnetic disturbance, and therefore that the entire theory of Light must be contained within the electromagnetic equations. One of the first consequences of this profound intuition is that the constant c (the ratio of the units of charge in the electromagnetic and electrostatic systems) is equal to the velocity of Light in a vacuum. Making use of the facts known to his age, Maxwell showed that this equality did approximately exist. More exact measurements of the two magni-tudes made by entirely independent methods have since provided

a continuous series of converging values, and with them a progressively strengthening confirmation of the fundamental notions
on which Maxwell built. Continuing his deductions, the great
physicist arrived at the famous ratio which still bears his name:
the ratio between the dielectric constant of a refractive medium
and its index of refraction. The facts, in the main, agree with it
very well; and it is well known that the electromagnetic Theory
of Light has since been confirmed by a great number of experiments. Hertz succeeded in producing invisible Light of very high
wave-length by purely electrical methods, and these Hertzian
waves, which have since given rise to wireless telegraphy and
telephony, have properties which conform to those of Light, if
allowance is made for the difference in wave-length. At the same
time the theory of the action of Light on Matter was developed
by the hypothesis that Matter consists of electrified centres on
which the electromagnetic fields of the light-waves act. This
development was due chiefly to Lorentz's work. The emission
and absorption of Light by Matter has similarly proved capable of
interpretation by the electromagnetic Theory, particularly in the
electronic form due to Lorentz—at any rate so far as was possible
while the existence of quanta was unknown. The discovery by
Zeeman in 1896 of the effect named after him (the influence of a
magnetic field on the lines emitted by a source placed within the
field) confirmed what Lorentz had predicted in striking fashion,
and gave further proof of the electromagnetic nature of Light.
The discovery of other magneto-optic and electro-optic phenomena
(the Kerr, Cotton and Mouton phenomena) have come since to
swell the number of such proofs. Down to the photo-electric effect,
with which we shall deal in a moment from another point of view,
every phenomenon gave a direct proof of the influence of Light on
electrified corpuscles, and thus gave fresh support to the idea that
there is a connection between Light and an electromagnetic field.
We are thus aware of a large group of facts designated electro-
optic phenomena.[1] Maxwell's Theory provides for all the essential

[1] Today this term denotes all optical phenomena revealing the electromagnetic character of Light.

elements of Fresnel's vectorial undulatory theory, and consequently it explains the neutral phenomena, specifically undulatory phenomena, vector phenomena (polarization) and electro-optic phenomena—in other words, the whole of the four classes of phenomena known some forty years ago. Thus for some time it rightly seemed to be the most complete imaginable synthesis of the facts connected with Light; which sufficiently explains both its well deserved fame and the occasionally somewhat despotic authority which it enjoyed. About 1900, then, the Theory of Light seemed to have been perfected. But a fifth group of phenomena, of a purely corpuscular character, was discovered: and all was once again uncertain.

The reader is already acquainted with these phenomena, whose unexpected discovery drew the attention of physicists back to the corpuscular view of Light. First and foremost among them is the photo-electric effect. I have previously dealt with this phenomenon, so that it will suffice to recall that in order to interpret the empirical laws, Einstein was led, about 1905, to the assumption that the energy of a light-ray of frequency v is divided into corpuscles of energy hv, where h is Planck's well known constant. This is the theory of light-quanta, as Einstein called them at the time, or of photons as we prefer to say today. His hypothesis accounts not only for the photo-electric effect, but also enables us to predict very precisely the Compton and Raman effects, which are phenomena discovered later on, and in which the corpuscular aspect of Light is manifested. The photon theory is a purely corpuscular theory of Light; and it provides a correct interpretation of the group of specifically corpuscular phenomena which I have just mentioned—the photo-electric effect, the Compton and Raman effects and others: the existence of a limit to the continuous spectrum of X-rays, the emission of Light in quanta in accordance with the Frequency Law so felicitously employed by Bohr to predict the spectral lines of atoms, etc. It also explains rectilinear propagation, reflection, radiation pressure, the Döppler effects and even Planck's Law, at any rate if it is assumed that Bose-Einstein statistics is to be applied to the photon. In short, the new corpuscular Theory of Light explains the totality of neutral phenomena and also those of a strictly corpuscular nature. On the other hand, it is completely baffled by such wave phenomena as interference and diffraction, by the "vectorial" phenomena of polarization and by electro-optic phenomena. It involves severe difficulties in connection with the coherence of wave series and the resolving power of optical instruments, as was shown by Lorentz shortly after the publication of Einstein's Treatise. These difficulties were removed only by the wholly novel and slightly revolutionary concepts on the localization of corpuscles in general involved by Wave Mechanics. Einstein's hypothesis, still further, may seem somewhat incoherent, since it defines the energy of the photon in terms of the frequency v of Light: this frequency, however, the notion of which it borrows from the theory of waves, it cannot define. To amend the concept of the photon, while retaining its useful elements, it was therefore necessary to effect a union between the idea of the wave and that of the corpuscle. It was Wave Mechanics which brought about this result.

* * *

Let us examine the contribution of Wave Mechanics to the Theory of Light. As we have seen in earlier contexts, the dual aspect of Light is revealed by the two groups of phenomena explained by the Theory of waves, and of corpuscles, respectively. Wave Mechanics accepted this dualism, and succeeded in establishing a general association between waves and corpuscles; an association within which the interconnection between photons and light-waves, as expressed by the relations of Einstein's Theory, is contained as a particular case. Further, a wholly novel idea was introduced, viz. that corpuscles have in general no determinate spatial position, but have only potential localizations obeying the laws of probability. In this way it proved possible to remove the objections raised by Lorentz to Einstein's Theory, and to provide an explanation of the phenomena of interference and diffraction which was at once satisfactory and compatible with the photon theory. In this way most of the difficulties which the new corpuscular view of Light seemed at first sight to involve were removed; an entirely novel merger with the undulatory conception was effected, with the net result that a striking synthesis was brought about. In this way Wave Mechanics succeeded, if not in constructing, at any rate in clearing the site for a comprehensive theory, explaining on the one hand the neutral and the corpuscular phenomena and on the other hand the scalar wave phenomena implied in Fresnel's theory of the light-variable. Nevertheless the synthesis is not a perfect one. For one thing it can be developed correctly (as I observed previously) only if a relativistic formulation of Wave Mechanics is adopted; and though there is such a formulation—which has actually been familiar since the beginnings of this new science—yet it is not one which today is considered really satisfactory. Still further, primitive Wave Mechanics, even though relativistic, introduces only a single wave-function to represent the undulatory properties of the corpuscle, namely the function Ψ which has a scalar character. Hence it permits no explanation of "vectorial" wave phenomena, nor of polarization. Finally, it affords us no means whatever of establishing a connection between a photon and an electromagnetic field, so that we would be enabled to retain the Maxwellian representation of a light-wave. In other words, Wave Mechanics has room only for half of Fresnel's work and none for that of Maxwell.

Having reached this point in our survey, let us set out in tabular form (page 156) the different theories of Light, at the same time linking them up with the five groups of luminous phenomena between which we have just distinguished. In the diagram the five groups of phenomena are shown in the following order: corpuscular phenomena, neutral phenomena, scalar wave phenomena, vectorial wave phenomena and electro-optic phenomena. The brackets show which are the various groups of phenomena interpreted by the different theories of Light.

The diagram seems to prove that at the moment there is no one theory to account for all luminous phenomena. Yet this inference is not wholly correct; for there are certain recent attempts not shown in the diagram (the Quantum Theory of the Field as

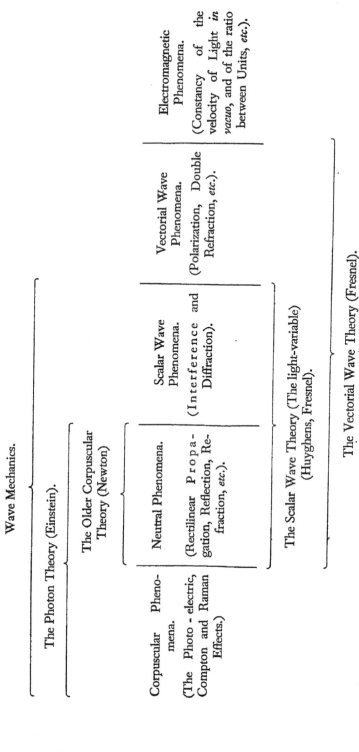

elaborated by Pauli and Heisenberg, Dirac's Theory of photons) which amount to an attempt to embrace every aspect of Light. Personally, however, I have never judged these attempts to be wholly satisfactory, despite their undoubted interest and despite the fact that some of the results to which they lead are of permanent value. My own belief is that the final synthesis of the known facts still remains to be made. I have made an attempt in this direction and, to conclude, I should like to set out the ideas which have guided me in this endeavour.

*   *

*

In our search after a theoretic synthesis, it would seem that we ought to be guided by the idea that there is a far-reaching symmetry between Matter and Light as far as their dual character (waves and corpuscles) is concerned. It was indeed from the notion of such a symmetry that Wave Mechanics started; the results to which it led are intellectually so satisfactory, and it is so truly the innermost reason of the success of the latest theories of Quantum Physics that, in my own opinion, it is indispensable. Yet it must be recognized that the dualistic theory of Light, having been the guide and model followed by the dualistic theory of Matter in the course of its development, is today lagging behind. By looking for the cause of this paradox, however, we shall be able to see in what direction we must attempt to advance. Among the first causes of these difficulties is the fact (already mentioned)[1] that, in the form under which its most rapid growth took place, Wave Mechanics is not relativistic, and also contains no element of symmetry which enables polarization to be included, nor any electromagnetic element permitting the definition of a Maxwellian light-wave. The velocity of the photon is equal to c, or at any rate so closely as to be indistinguishable from it; and in the Theory of Relativity c is the upper limit of the velocities of corpuscles: so that, quite clearly, Relativity Mechanics should be used in dealing with the photon, while Wave Mechanics can succeed in describing the photon only if it is used in

[1] cf. p. 130.

a relativistic form. Consequently, so far as Light is concerned, it is necessary to formulate a Wave Mechanics having a place for Relativity.

Another cause for the difficulties met with in erecting a Theory of photons, having some symmetry with the Wave Theory of the elementary corpuscles of Matter, is that photons have properties which clearly differentiate them from such material corpuscles as the electron. First, every large group of photons obeys Bose-Einstein statistics, while electrons, on the contrary, obey Fermi-Dirac statistics, and though we are still ignorant of the ultimate physical conditions which decide why the different kinds of particles follow one kind of statistics rather than the other, we still feel that the photon and the electron, since they behave so differently in statistical phenomena, cannot be of an altogether similar character. Secondly, in the photo-electric effect the photon disappears and is apparently annihilated, whereas there is no corresponding phenomenon in the case of material corpuscles, unless we except the recently discovered phenomenon of the dematerialization of pairs of electrons having opposite signs. I feel, accordingly, that if we wish to form a theory of the photon, the first necessity is to make use of a relativistic form of Wave Mechanics, and that (secondly) we must introduce "something extra" to differentiate photons from elementary material corpuscles.

The first part of this programme is immediately fulfilled if we substitute the more elaborate form provided by Dirac, in his Theory of the magnetic electron,[1] for the primitive form of Wave Mechanics. Dirac's Theory is, in fact, compatible with Relativity at least in so far as it is applicable to corpuscles having every velocity up to the limit, c; and more particularly, it assigns to the corpuscles energy and momentum in agreement with the formulae of Relativity, which is essential if we are to utilize the fundamental Einsteinian formulae of the photon Theory which are requisite for the interpretation of the photo-electric effect and the Compton effect. There is also the very striking fact that Dirac's Theory automatically introduces an element of symmetry, in the form of

[1] P. 146.

spin, which reveals a definite kinship with polarization, and further permits of the determination of the electromagnetic magnitudes, magnetic moment and electric moment, associated with the corpuscle. From all this it is clear that Dirac's Theory is a considerable step towards a *rapprochement* between Wave Mechanics (which applies to material elements) and the synthetic theory which we desiderate for the interpretation of *all* the properties of Light. At the same time it is not enough that the photon shall simply be a corpuscle whose mass is zero or negligible, and which obeys the equations of Dirac's Theory. Actually this Theory, attractive as it is in its simplicity, leads to a model of the photon having, so to put it, only half the symmetry of the real photon. The photon, still further, ought certainly to obey Fermi statistics, and should also be incapable of annihilation in the photo-electric effect. Hence the "something extra" ought undoubtedly to be added.

In my search for this "something extra," I have found that the only phenomenon known to us today in which material corpuscles vanish, in a way analogous to that in which the photon disappears in the photo-electric effect, is that in which pairs of electrons of opposite signs are dematerialized. The positive electron was discovered only a few years ago and its discovery was a striking confirmation of certain features in Dirac's equations which up to that point had rather seemed to be blemishes. Suitably interpreted, then, Dirac's Theory shows that the existence of a negative electron ought to imply the existence of a positive anti-electron. Still more generally, there ought to be an anti-corpuscle corresponding to every corpuscle obeying Dirac's equations, and standing to the latter in the same relation as the positive electron to the negative electron. These predictions were confirmed by the discovery of the positive electron, so that it has become tempting to imagine the photon as consisting of a corpuscle of negligible mass and charge obeying Dirac's equations, and associated with an anti-corpuscle of the same character. It is an attractive hypothesis, and from the mathematical point of view it can be completely worked out. It is easy to understand how a photon constructed in this way could be annihilated in the presence of Matter by transferring to it the whole of its energy, a process analogous to the annihilation of a pair of electrons in the phenomenon of dematerialization. This annihilation—a quantum transition—would then constitute the photo-electric effect, the fundamental importance of which from the theoretical point of view has already been pointed out; and it ought then to be possible to define the electro-magnetic field as a function of this transition. Actually, indeed, it can be shown that it is possible to connect with this transitional process of annihilation an electromagnetic field completely identical with that which defines Maxwell's wave—an extremely interesting fact. Further, since the photon is thus assumed to consist of a corpuscle and an anti-corpuscle, both of them possessing spin, it should now follow Bose-Einstein statistics.

It should be observed that Jordan has developed a kind of variant on this theory, differing from it in certain essential particulars, especially in treating the photon as a mere appearance and not as a genuine unit. This view has been further developed recently by de Kronig; it has certain interesting aspects, but is not finished enough to allow it to be judged as a whole.

*   *
*

Whatever may be the position of these, the most recent, theories with regard to their details, it would seem that they constitute a road leading us towards that synthetic theory which, in the diagram, would unite the explanations of the five distinct groups of phenomena connected with Light which we have been discussing, within a single all-comprehensive bracket. It is the hope, ever re-born, of theorists in Physics, that despite the continually increasing complexity of the known phenomena they may succeed in constructing more and more comprehensive synthetic theories, each of them embracing and completing all that went before. The history of the different theories of Light has given us a splendid example of the success of such a series of syntheses in one particular branch of Physics. We must not be surprised, however, if on many occasions the discovery of a new series of phenomena destroys our finest theories like a house of cards; for the richness of Nature is always greater than our imagination. It requires, indeed, some boldness in physicists to attempt the reconstruction by thought of part of the plan of the Universe: the miracle is that they have sometimes succeeded.

IV

WAVE MECHANICS

I

THE UNDULATORY ASPECTS OF THE ELECTRON [1]

WHEN, in 1920, I resumed my investigations in theoretical Physics after a long interruption through circumstances out of my own control, I was far from imagining that this research would within a few years be rewarded by the lofty and coveted distinction given each year by the Swedish Academy of Sciences: the Nobel Prize in Physics. At that time what drew me towards theoretical Physics was not the hope that so high a distinction would ever crown my labours: what attracted me was the mystery which was coming to envelop more and more deeply the structure of Matter and of radiation in proportion as the strange concept of the quantum, introduced by Planck about 1900 during his research on black body radiation, came to extend over the entire field of Physics.

But to explain the way in which my research came to develop I must first outline the critical period through which Physics had for the last twenty years been passing.

*   *   *

Physicists had for long been wondering whether Light did not consist of minute corpuscles in rapid motion, an idea going back to the philosophers of antiquity, and sustained in the eighteenth century by Newton. After interference phenomena had been discovered by Thomas Young, however, and Augustin Fresnel had completed his important investigation, the assumption that Light had a granular structure was entirely disregarded, and the

[1] Address delivered at Stockholm on receiving the Nobel Prize, December 12, 1929.

Wave Theory was unanimously adopted. In this way the physicists of last century came to abandon completely the idea that Light had an atomic structure. But the Atomic Theory, being thus banished from optics, began to achieve great success, not only in Chemistry, where it provided a simple explanation of the laws of definite proportions, but also in pure Physics, where it enabled a fair number of the properties of solids, liquids and gases to be interpreted. Among other things it allowed the great kinetic theory of gases to be formulated, which, in the generalized form of statistical Mechanics, has enabled clear significance to be given to the abstract concepts of thermodynamics. We have seen how decisive evidence in favour of the atomic structure of electricity was also provided by experiments. Thanks to Sir J. J. Thomson, the notion of the corpuscle of electricity was introduced; and the way in which H. A. Lorentz has exploited this idea in his electron Theory is well known.

Some thirty years ago, then, Physics was divided into two camps. On the one hand there was the Physics of Matter, based on the concepts of corpuscles and atoms which were assumed to obey the classical laws of Newtonian Mechanics; on the other hand there was the Physics of radiation, based on the idea of wave propagation in a hypothetical continuous medium: the ether of Light and of electromagnetism. But these two systems of Physics could not remain alien to each other: an amalgamation had to be effected; and this was done by means of a theory of the exchange of energy between Matter and radiation. It was at this point, however, that the difficulties began; for in the attempt to render the two systems of Physics compatible with each other, incorrect and even impossible conclusions were reached with regard to the energy equilibrium between Matter and radiation in an enclosed and thermally isolated region: some investigators even going so far as to say that Matter would transfer all its energy to radiation, and hence tend towards the temperature of absolute zero. This absurd conclusion had to be avoided at all costs; and by a brilliant piece of intuition Planck succeeded in doing so. Instead of assuming, as did the classical Wave Theory, that a light-source emits its radiation continuously, he assumed that it emits it in equal and finite quantities—in quanta. The energy of each quantum, still further, was supposed to be proportional to the frequency of the radiation, v, and to be equal to hv, where h is the universal constant since known as Planck's Constant.

The success of Planck's ideas brought with it some serious consequences. For if Light is emitted in quanta, then surely, once radiated, it ought to have a granular structure. Consequently the existence of quanta of radiation brings us back to the corpuscular conception of Light. On the other hand, it can be shown—as has in fact been done by Jeans and H. Poincaré—that if the motion of the material particles in a light-source obeyed the laws of classical Mechanics, we could never obtain the correct Law of black body radiation—Planck's Law. It must therefore be admitted that the older dynamics, even as modified by Einstein's Theory of Relativity, cannot explain motion on a very minute scale.

The existence of a corpuscular structure of Light and of other types of radiation has been confirmed by the discovery of the photo-electric effect which, as I have already observed, is easily explained by the assumption that the radiation consists of quanta—hv—capable of transferring their entire energy to an electron in the irradiated substance; and in this way we are brought to the theory of light-quanta which, as we have seen, was advanced in 1905 by Einstein—a theory which amounts to a return to Newton's corpuscular hypothesis, supplemented by the proportionality subsisting between the energy of the corpuscles and the frequency. A number of arguments were adduced by Einstein in support of his view, which was confirmed by Compton's discovery in 1922 of the scattering of X-rays, a phenomenon named after him. At the same time it still remained necessary to retain the Wave Theory to explain the phenomena of diffraction and interference, and no means was apparent to reconcile this Theory with the existence of light-corpuscles.

I have pointed out that in the course of investigation some doubt had been thrown on the validity of small-scale Mechanics. Let us imagine a material point describing a small closed orbit—

an orbit returning on itself; then according to classical dynamics there is an infinity of possible movements of this type in accordance with the initial conditions, and the possible values of the energy of the moving material point form a continuous series. Planck, on the other hand, was compelled to assume that only certain privileged movements—*quantized* motion—are possible, or at any rate stable, so that the available values of the energy form a discontinuous series. At first this seemed a very strange idea; soon, however, its truth had to be admitted, because it was by its means that Planck arrived at the correct Law of black body radiation and because its usefulness has since been proved in many other spheres. Finally, Bohr founded his famous atomic Theory on this idea of the quantization of atomic motion—a theory so familiar to scientists that I will refrain from summing it up here.

Thus we see once again it had become necessary to assume two contradictory theories of Light, in terms of waves, and of corpuscles, respectively; while it was impossible to understand why, among the infinite number of paths which an electron ought to be able to follow in the atom according to classical ideas, there was only a restricted number which it could pursue in fact. Such were the problems facing physicists at the time when I returned to my studies.

*   *   *

When I began to consider these difficulties I was chiefly struck by two facts. On the one hand the Quantum Theory of Light cannot be considered satisfactory, since it defines the energy of a light-corpuscle by the equation $W = h\nu$, containing the frequency ν. Now a purely corpuscular theory contains nothing that enables us to define a frequency; for this reason alone, therefore, we are compelled, in the case of Light, to introduce the idea of a corpuscle and that of periodicity simultaneously.

On the other hand, determination of the stable motion of electrons in the atom introduces integers; and up to this point the only phenomena involving integers in Physics were those of interference and of normal modes of vibration. This fact suggested to me the idea that electrons too could not be regarded simply as corpuscles, but that periodicity must be assigned to them also.

In this way, then, I obtained the following general idea, in accordance with which I pursued my investigations:—that it is necessary in the case of Matter, as well as of radiation generally and of Light in particular, to introduce the idea of the corpuscle and of the wave simultaneously: or in other words, in the one case as well as in the other, we must assume the existence of corpuscles accompanied by waves. But corpuscles and waves cannot be independent of each other: in Bohr's terms, they are two complementary aspects of Reality: and it must consequently be possible to establish a certain parallelism between the motion of a corpuscle and the propagation of its associated wave. The first object at which to aim, therefore, was to establish the existence of this parallelism.

With this in view, I began by considering the simplest case: that of an isolated corpuscle, i.e. one removed from all external influence; with this we wish to associate a wave. Let us therefore consider first of all a reference system $O \, x_0 y_0 z_0$ in which the corpuscle is at rest: this is the "proper" system for the corpuscle according to the Theory of Relativity. Within such a system the wave will be stationary, since the corpuscle is at rest; its phase will be the same at every point, and it will be represented by an expression of the form $\sin 2\pi\nu_0(t_0 - \tau_0)$, t_0 being the "proper" time of the corpuscle, and τ_0 a constant.

According to the principle of inertia the corpuscle will be in uniform rectilinear motion in every Galilean system. Let us consider such a Galilean system, and let v be the velocity of the corpuscle in this system. Without loss of generality, we may take the direction of motion to be the axis of x. According to the Lorentz transformation, the time t employed by an observer in this new system is linked with the proper time t_0 by the relation:

$$t_0 = \frac{t - \dfrac{\beta x}{c}}{\sqrt{1 - \beta^2}}$$

where $\beta = v/c$.

Hence for such an observer the phase of the wave will be given by

$$\sin 2\pi \frac{\nu_0}{\sqrt{1-\beta^2}} \left(t - \frac{\beta x}{c} - \tau_0 \right).$$

Consequently the wave will have for him a frequency

$$\nu = \frac{\nu_0}{\sqrt{1-\beta^2}}$$

and will move along the axis of x with the phase-velocity

$$V = \frac{c}{\beta} = \frac{c^2}{v}.$$

If we eliminate β from the two preceding formulae we shall easily find the following relation, which gives the index of refraction of free space, n, for the waves under consideration

$$n = \sqrt{1 - \frac{\nu_0^2}{\nu^2}}$$

To this "law of dispersion" there corresponds a "group velocity." You are aware that the group velocity is the velocity with which the resultant amplitude of a group of waves, with almost equal frequencies, is propagated. Lord Rayleigh has shown that this velocity U satisfies the equation

$$\frac{1}{U} = \frac{1}{c}\frac{d\,(n\nu)}{d\nu}$$

Here we find that $U = v$, which means that the velocity of the group of waves in the system $x\,y\,\zeta\,t$ is equal to the velocity of the corpuscle in this system. This relation is of the greatest importance for the development of the Theory.

Accordingly, in the system $x\,y\,\zeta\,t$ the corpuscle is defined by the frequency v and by the phase-velocity V of its associated wave. In order to establish the parallelism mentioned above, we must try to connect these magnitudes to the mechanical magnitudes—to energy and momentum. The ratio between energy and frequency is one of the most characteristic relations of the Quantum Theory; and since, still further, energy and frequency are transformed when the Galilean system of reference is changed, it is natural to establish the equation

$$\text{Energy} = h \times \text{frequency, or } W = h\nu$$

where h is Planck's constant. This relation must apply to all Galilean systems; and in the proper system of the corpuscle where, according to Einstein, the energy of the corpuscle is reduced to its internal energy $m_0 c^2$ (where m_0 is its proper mass) we have $h\nu_0 = m_0 c^2$.

This relation gives the frequency ν_0 as a function of the proper mass m_0, or inversely.

The momentum is a vector $\mathbf{p}$ equal to $\dfrac{m_0 \mathbf{v}}{\sqrt{1 - \beta^2}}$, where $|\mathbf{v}| = v$, and then we have

$$p = |\mathbf{p}| = \frac{m_0 v}{\sqrt{1 - \beta^2}} = \frac{W v}{c^2} = \frac{h\nu}{V} = \frac{h}{\lambda}.$$

The quantity λ is the wave-length—the distance between two consecutive wave-crests

hence
$$\lambda = \frac{h}{p}.$$

This is a fundamental relation of the Theory.

* * *

All that has been said refers to the very simple case where there is no field of force acting on the corpuscle. I shall now indicate very briefly how the Theory can be generalized for the case of a corpuscle moving in a field of force not varying with time derived from a potential energy function $F(x, y, z)$. Arguments into which I shall not enter here lead us in such a case to assume that the propagation of the wave corresponds to an index of refraction varying from point to point in space in accordance with the formula

$$n(x, y, z) = \sqrt{\left[1 - \frac{F(x, y, z)}{h\nu}\right]^2 - \frac{\nu_0^2}{\nu^2}}$$

or, as a first approximation, if we neglect the corrections introduced by the Theory of Relativity

$$n(x, y, z) = \sqrt{\frac{2(E - F)}{m_0 c^2}} \quad \text{with} \quad E = W - m_0 c^2.$$

The constant energy W of the corpuscle is further connected with the constant frequency ν of the wave by the relation

$$W = h\nu$$

while the wave-length λ, which varies from one point to the other in the field of force, is connected with the momentum p (which is also variable) by the relation

$$\lambda(x, y, z) = \frac{h}{p(x, y, z)}.$$

Here again we show that the velocity of the wave-group is equal to the velocity of the corpuscle. The parallelism thus established between the corpuscle and its wave enables us to identify Fermat's Principle in the case of waves and the Principle of Least Action in that of corpuscles, for constant fields. Fermat's Principle states that the ray in the optical sense passing between two points A and B in a medium whose index is $n(x, y, z)$, variable from one point to the other but constant in time, is such that the integral $\int_A^B n \, dl$, taken along this ray, shall be an *extremum*. On the other hand, Maupertuis' Principle of Least Action asserts that the trajectory of a corpuscle passing through two points A and B is such that the integral $\int_A^B p \, dl$ taken along the trajectory shall be an *extremum*, it being understood that we are considering only the motion corresponding to a given value of energy. According to the relations already established between the mechanical and the wave magnitudes, we have

$$n = \frac{c}{V} = \frac{c}{\nu} \cdot \frac{1}{\lambda} = \frac{c}{h\nu} \cdot \frac{h}{\lambda} = \frac{c}{W} p = \text{constant} \times p$$

since W is constant in a constant field. Hence it follows that Fermat's Principle and Maupertuis' Principle are each a rendering of the other: the possible trajectories of the corpuscle are identical with the possible rays of its wave.

These ideas lead to an interpretation of the conditions of stability introduced by the Quantum Theory. If we consider a closed trajectory C a constant field it is quite natural to assume that the phase of the associated wave should be a uniform function along this trajectory. This leads us to write

$$\int_c \frac{dl}{\lambda} = \int_c \frac{1}{h} p \, dl = \text{an integer.}$$

Now this is exactly the condition of the stability of atomic periodic motion, according to Planck. Thus the quantum conditions of stability appear as analogous to resonance phenomena, and the appearance of integers here becomes as natural as in the theory of vibrating cords and discs.

* * *

The general formulae establishing the parallelism between waves and corpuscles can be applied to light-corpuscles if we assume that in that case the rest-mass m_0 is infinitely small. If then for any given value of the energy W we make m_0 tend to zero, we find that both v and V tend to c, and in the limit we obtain the two fundamental formulae on which Einstein erected his Theory of Light-quanta

$$W = h\nu \quad p = \frac{h\nu}{c}.$$

Such were the principal ideas which I had developed during my earlier researches. They showed clearly that it was possible to establish a correspondence between waves and corpuscles of such a kind that the Laws of Mechanics correspond to those of geometrical optics. But we know that in the Wave Theory geometrical optics is only an approximation: there are limits to the validity of this approximation, and especially when the phenomena of interference and of diffraction are concerned it is wholly inadequate. This suggests the idea that the older Mechanics too may be no more than an approximation as compared with a more comprehensive Mechanics of an undulatory character. This was what I expressed at the beginning of my researches when I said that a new Mechanics must be formulated, standing in the same relation to the older Mechanics as that in which wave optics stands to geometrical optics. This new Mechanics has since been developed, thanks in particular to the fine work done by Schrödinger. It starts from the equations of wave propagation, which are taken as the basis, and rigorously determines the temporal changes of the wave associated with a corpuscle. More particularly, it has succeeded in giving a new and more satisfactory form to the conditions governing the quantization of intra-atomic motion: for, as we have seen, the older conditions of quantization are encountered again if we apply geometrical optics to the waves associated with intra-atomic corpuscles; and there is strictly no justification for this application.

I cannot here trace even briefly the development of the new Mechanics. All that I wish to say is that on examination it has shown itself to be identical with a Mechanics developed independently, first by Heisenberg and later by Born, Jordan, Pauli, Dirac and others. This latter Mechanics—Quantum Mechanics— and Wave Mechanics are, from the mathematical point of view, equivalent to each other.

Here we must confine ourselves to a general consideration of the results obtained. To sum up the significance of Wave Mechanics, we can say that a wave must be associated with each particle, and that a study of the propagation of the wave alone can tell us anything about the successive localizations of the corpuscle in space. In the usual large-scale mechanical phenomena, the localizations predicted lie along a curve which is the trajectory in the classical sense of the term. What, however, happens if the wave is not propagated according to the laws of geometrical optics; if, for example, interference or diffraction occurs? In such a case we can no longer assign to the corpuscle motion in accordance with classical dynamics. So much is certain. But a further question arises: Can we even suppose that at any given moment the corpuscle has an exactly determined position within the wave, and that in the course of its propagation the wave carries the corpuscle with it, as a wave of water would carry a cork? These are difficult questions, and their discussion would carry us too far and actually to the borderland of Philosophy. All that I shall say here is that the general modern tendency is to assume that it is not always possible to assign an exactly defined position within the wave to the corpuscle, that whenever an observation is made enabling us to localize the corpuscle, we are invariably led to attribute to it a position inside the wave, and that the probability that this position is at a given point, M, within the wave is proportional to the square of the amplitude, or the intensity, at M.

What has just been said can also be expressed in the following way. If we take a cloud of corpuscles all associated with the same wave, then the intensity of the wave at any given point is proportional to the density of the cloud of corpuscles at that point, i.e. to the number of corpuscles per unit of volume around that point. This assumption must be made in order to explain how it is that in the case of interference the luminous energy is found concentrated at those points where the intensity of the wave is at a maximum: if it is assumed that the luminous energy is transferred by light-corpuscles, or photons, then it follows that the density of the photons in the wave is proportional to this intensity.

This rule by itself enables us to understand the way in which the undulatory theory of the electron has been verified experimentally.

For let us imagine an indefinite cloud of electrons, all having the same velocity and moving in the same direction. According to the fundamental ideas of Wave Mechanics, we must associate with this cloud an infinite plane wave having the form

$$a \exp. 2\pi i \left[\frac{W}{h} t - \frac{\alpha x + \beta y + \gamma z}{\lambda} \right]$$

where α, β, γ are the direction cosines of the direction of propagation, and where the wave-length λ is equal to $\frac{h}{p}$. If the electrons have no extremely high velocity, we may say

$$p = m_0 v$$

and hence

$$\lambda = \frac{h}{m_0 v}$$

m_0 being the rest-mass of the electron.

In practice, to obtain electrons having the same velocity they are subjected to the same potential difference P. We then have

$$\frac{1}{2} m_0 v^2 = e\,P$$

Consequently

$$\lambda = \frac{h}{\sqrt{2 m_0\, e\, P}}$$

Numerically, this gives

$$\lambda = \frac{12\cdot 24}{\sqrt{P}} \cdot 10^{-8} \text{ cm. } (P \text{ in volts}).$$

As we can only use electrons that have fallen through a potential difference of at least some tens of volts, it follows that the wave-length λ, assumed by the Theory, is at most of the order of 10^{-8} cm., i.e. of the order of the Ångström unit.[1] This is also the order of magnitude of the wave-lengths of X-rays.

The length of the electron wave being thus of the same order as that of X-rays, we may fairly expect to be able to obtain a scattering of this wave by crystals, in complete analogy to the Laue phenomenon in which, in a natural crystal like rock salt, the atoms of the substances composing the crystal are arranged at regular intervals of the order of one Ångström, and thus act as scattering centres for the waves. If a wave having a length of one Ångström encounters the crystal, then the waves scattered agree

[1] [This is one ten-millionth mm.]

in phase in certain definite directions. In these directions the total intensity scattered exhibits a strong maximum. The location of these maxima of scattering is given by the well known mathematical Theory elaborated by Laue and Bragg, which gives the position of the maxima in terms of the distance between the atomic arrangements in the crystal and of the length of the incident wave. For X-rays the Theory has been triumphantly substantiated by Laue, Friedrich and Knipping, and today the diffraction of X-rays by crystals has become a quite commonplace experiment. The exact measurement of the wave-lengths of X-rays is based on this diffraction, as I need hardly recall in a country where Siegbahn and his collaborators are pursuing their successful labours.

In the case of X-rays, the phenomenon of diffraction by crystals was a natural consequence of the idea that these rays are undulations analogous to Light, and differ from Light only by their shorter wave-length. But for electrons no such view could be entertained, so long as the latter were looked upon as being merely minute corpuscles. If, on the other hand, we assume that the electron is associated with a wave, and that the density of a cloud of electrons is measured by the intensity of the associated wave, we may then expect that there will be effects in the case of electrons similar to the Laue effect. In that event, the electron wave will be scattered with an intensity in certain directions which the Laue-Bragg Theory enables us to calculate, on the assumption that the wave-length is $\lambda = \dfrac{h}{mv}$, a length corresponding to the known velocity v of the electrons falling on the crystal. According to our general principle, the intensity of the scattered wave measures the density of the cloud of scattered electrons, so that we may expect to find large numbers of scattered electrons in the directions of the maxima. If this effect actually occurs, it would provide a crucial experimental proof of the existence of a wave associated with the electron, its length being $\dfrac{h}{mv}$. In this way the fundamental idea of Wave Mechanics would be provided with a firm experimental foundation.

Now experiment—which is the last Court of Appeal of theories

—has shown that the diffraction of electrons by crystals actually occurs, and that it follows the Laws of Wave Mechanics exactly and quantitatively. It is (as we have seen already) to Davisson and Germer, working at the Bell Laboratories in New York, that the credit belongs of having been the first to observe this phenomenon by a method similar to that used by Laue for X-rays. Following up the same experiments, but substituting for the single crystal a crystalline powder, in accordance with the method introduced for X-rays by Debye and Scherrer, Professor G. P. Thomson, of Aberdeen, the son of the great Cambridge physicist, Sir J. J. Thomson, has discovered the same phenomena. At a later stage Rupp in Germany, Kikuchi in Japan and Ponte in France have also reproduced them under varying experimental conditions. Today the existence of the effect is no longer subject to doubt, and the minor difficulties of interpretation which Davisson's and Germer's earlier experiments had raised have been resolved in a satisfactory manner. Rupp has actually succeeded in obtaining the diffraction of electrons in a particularly striking form. A grating is employed— a metal or glass surface, either plane or slightly curved, on which equidistant lines have been mechanically drawn, the interval between them being of an order of magnitude comparable to that of the wave-lengths of Light. Between the waves diffracted by these lines there will be interference, and the interference will give rise to maxima of diffracted Light in certain directions depending on the distance between the lines, on the direction of the Light falling on the grating and on the wave-lengths. For a long time it remained impossible to obtain similar effects with gratings of this kind produced by human workmanship when X-rays were used instead of Light. The reason for this was that the wave-length of X-rays is a great deal shorter than that of Light, and that there is no instrument capable of drawing lines on any surface at intervals of the order of X-ray wave-lengths. Ingenious physicists, however, (Compton and Thibaud) succeeded in overcoming the difficulty. Let us take an ordinary optical grating and let us look at it more or less at a tangent. The lines of the grating will then seem to be much closer together than they actually are. For X-rays falling on the grating at this grazing angle, the conditions will be the same as though the lines were extremely close together, and diffraction effects like those of Light will be produced. The physicists just mentioned have proved that such was in fact the case. But now—since the electron wave-lengths are of the same order as those of X-rays—we should also be able to obtain these diffraction phenomena by causing a beam of electrons to fall on such an optical grating at a very small grazing angle. Rupp succeeded in doing this. He was thus enabled to measure the length of electron waves by comparing it directly with the distance between the lines drawn mechanically on the grating.

*   *
*

We thus find that in order to describe the properties of Matter, as well as those of Light, we must employ waves and corpuscles simultaneously. We can no longer imagine the electron as being just a minute corpuscle of electricity: we must associate a wave with it. And this wave is not just a fiction: its length can be measured and its interferences calculated in advance. In fact, a whole group of phenomena was in this way predicted before being actually discovered. It is, therefore, on this idea of the dualism in Nature between waves and corpuscles, expressed in a more or less abstract form, that the entire recent development of theoretical Physics has been built up, and that its immediate future development appears likely to be erected.

2

WAVE MECHANICS AND ITS INTERPRETATIONS [1]

DURING the last ten years the new Wave Mechanics has received strong experimental confirmation, thanks to the discovery of an interesting phenomenon entirely unknown before—the diffraction of electrons by crystals.

From one point of view we may say that this discovery is the exact counterpart of the earlier discovery of the photo-electric effect: for it shows us that, for Matter as well as for Light, we had hitherto been neglecting one of the aspects of the real physical world.

The photo-electric effect has shown us that the Wave Theory of Light, erected on a firm foundation by Fresnel and later imported into electromagnetic theory by Maxwell, contains an important part of the truth; but it is not wholly adequate, and in a certain sense it is necessary to revert to the idea of light-corpuscles originally suggested by Newton.

I have previously remarked that in his famous Theory of black-body radiation, Planck had been led to assume that radiation of frequency v is invariably emitted and absorbed in equal and finite quantities—in quanta whose value was hv where h was the constant to which Planck's name will remain attached. To interpret the photo-electric effect (as has been seen) Einstein had only to advance the hypothesis—a hypothesis entirely compatible with Planck's ideas—that Light consists of corpuscles, and that the energy of the corpuscles in Light of frequency v is hv. When a light-corpuscle in traversing Matter meets with an electron at rest, it can transfer to it its energy hv. The electron is in this way

[1] Delivered at the Meeting of The British Association in Glasgow, 1928.

set in motion and it will leave the Matter with kinetic energy equal to the difference between the energy $h\nu$ which it has received and the work requisite to enable it to leave the Matter. Now this is precisely the experimental Law of the photo-electric effect in the form in which it has been shown to apply to all radiation from the ultra-violet to X- and γ-rays.

Developing his idea, Einstein has shown that if the theory of light-corpuscles, or light-quanta, is accepted, we must attribute to each of these corpuscles a momentum equal to $p = \dfrac{h\nu}{c}$ together with the energy $W = h\nu$.[1] These two relations define the corpuscle of Light with frequency ν in terms of energy.

More recently Einstein's corpuscular Theory has been confirmed by the discovery of the Compton effect. This consists in the fact that a beam of X-rays falling on Matter has its frequency lowered while at the same time electrons are set in motion more or less rapidly. The phenomenon is easily interpreted by the assumption that there is an encounter or collision between a light-corpuscle and an electron originally at rest in the Matter in question. During the collision the electron borrows energy from the light-corpuscle and begins to move. The corpuscle has thus lost part of its energy, and since the relation $W = h\nu$ must always be valid, the frequency of the light-quantum will be lowered after the collision. The theory of the Compton effect, based on the two equations $W = h\nu$ and $p = \dfrac{h\nu}{c}$, has been developed by Compton himself and by Debye. Quantitatively it has been confirmed by experiment—and another success has been won for the theory that Light has a corpuscular structure.

But despite this success (as I have observed in the previous passage referred to) the Theory of light-quanta itself is not quite satisfactory. First of all, the group of diffraction and interference phenomena demands the introduction of the concept of waves; further, the two fundamental equations $W = h\nu$ and $p = \dfrac{h\nu}{c}$

---

[1] cf. p. 168.

imply the existence of a frequency ν. This is quite enough to prove that Light cannot consist simply of moving corpuscles. At the same time the discovery of the photo-electric effect and its confirmation by the Compton effect have shown that we must introduce into optics the notion of a corpuscle side by side with that of a wave. Thus it appears that a curious dualism in Nature has been revealed.

But if the corpuscular aspect had been unduly neglected in the Theory of Light during the last hundred years in favour of the wave aspect, may we not fairly ask whether the converse error has been committed in the Theory of Matter? Is it not the case that the wave aspect has been unduly neglected, and that thought has been unduly concentrated on the corpuscular aspect? Such were the questions which I asked myself some years ago, when I began to consider the analogy between the Principle of Least Action and Fermat's Principle, and the mysterious quantum conditions introduced into intra-atomic dynamics by Planck, Bohr, Wilson and Sommerfeld. By a course of argument which I will here pass over we can reach the conviction that it is essential to introduce waves into the Theory of Matter—and to do so in the following form.

Let there be a material particle, for example an electron, of mass m moving freely with constant velocity v. If we take the expressions provided by the Theory of Relativity, its energy and its momentum are

$$W = \frac{mc^2}{\sqrt{1 - \beta^2}} \quad \mathbf{p} = \frac{m\mathbf{v}}{\sqrt{1 - \beta^2}} = \frac{W}{c^2}\mathbf{v} \quad \left(\beta = \frac{|\mathbf{v}|}{c}\right) \quad (1)$$

where c is the velocity of Light in empty space. According to the new conception we must associate a wave with this particle, and this wave must be propagated in the direction of motion, its frequency being

$$\nu = \frac{W}{h} \tag{2}$$

and its phase-velocity

$$V = \frac{c^2}{v} = \frac{c}{\beta} \tag{3}$$

We thus have

$$\frac{h\nu}{V} = \frac{W\nu}{c^2} = p \qquad (4)$$

and consequently, if λ is the wave-length of the associated wave,

$$\lambda = \frac{V}{\nu} = \frac{h}{p}. \qquad (5)$$

If we try to apply these formulae to a light-corpuscle instead of to a material particle, then $v = c$, in which case we find

$$W = h\nu \qquad p = \frac{h\nu}{c}. \qquad (6)$$

Now these are the fundamental formulae of the Theory of light-quanta. Consequently our formulae (2) to (5) are generalized: they are applicable equally to Matter and to radiation and they express the fact that in each case alike it is necessary to introduce the ideas of the corpuscle and the wave side by side.

The older Mechanics corresponds to the case where the propagation of the associated wave agrees with the laws of geometrical optics, as follows with special clearness from Schrödinger's excellent Papers. In this case the corpuscle can be regarded as travelling along one of the "rays" of the wave with a velocity equal to the group velocity as ascertained by Lord Rayleigh. Under these conditions, accordingly, we may treat the corpuscle as consisting of a group of waves having nearly identical frequencies. From this we obtain a physical schema of the corpuscle which would be satisfactory enough if it were possible to generalize it. Unfortunately this is not the case.

It should be noted that if the associated wave moves in accordance with the laws of geometrical optics, there can be no experiment to prove the existence of associated waves; for in such a case the result of any experiment whatever can only be regarded as a demonstration of the correctness of the laws of the older Mechanics. But the position is altered when the conditions in which the associated wave is propagated are such that the approximations of geometrical optics no longer suffice to account for them. In such a case the new concepts would lead us to expect that we should be able to observe the phenomena which the older systems of Mechanics were entirely unable to predict, and which are characteristic of the new undulatory conception of dynamics.

Within the field of the new dynamics, still further, the principle which appears to be best established is that the square of the amplitude of the wave—its intensity—must, at any instant and at any given point, measure the probability that the associated corpuscle is at that point at that moment. A little reflection shows that this principle is required to account for the phenomena of interference and diffraction of Light, since it is the case in optics that wherever a Fresnel wave has maximum intensity, there we shall find on an average the maximum luminous energy. Our guiding notion, therefore, is to effect the closest possible union between the Theory of Light and that of Matter, so that it is quite natural to extend the principle necessary in the case of Light to the case of material particles.

We thus come to the conclusion that material corpuscles ought to exhibit phenomena analogous to those of interference and diffraction in the sphere of Light, and that the methods of calculation ought to be much the same in each case. According to the new views, then, a cloud of electrons having the same velocity ought to be associated with a plane monochromatic wave. Let us now assume that this cloud of electrons falls on a medium having a regular structure, like a crystal. If the distances separating the elementary constituents of this structure are of the same order of magnitude as the length of the incident wave the wave will be diffracted, and the resulting scattered wave will have a maximum amplitude in certain easily calculable directions. Accordingly we may expect to find that after scattering has taken place the electrons will be concentrated in certain directions. In this case we should have found the exact counterpart of Laue's experiments with X-rays; and if the results agree with theoretical predictions we shall have obtained an immediate and powerful proof that it is necessary to add the concept of a wave to that of the corpuscle even in the case of Matter.

I observed in the preceding Chapter that by different methods and under different conditions experiments of such a tenour were made by Davisson and Germer in New York, by Professor G. P. Thomson at Aberdeen and by Rupp at Göttingen. Theory and experiment agreed excellently, and the discrepancy found in Davisson and Germer's first experiment seemed to be easily explained if the index of refraction of the waves in the crystal was taken into account. We found, too, that Rupp even succeeded in obtaining the diffraction of a beam of electrons by using an ordinary grating with electrons falling on it at a tangent. Experiment had thus confirmed theory as fully as could be expected.

*   *   *

Thus a brilliant series of experiments (to repeat) had demonstrated that corpuscles and waves must be brought into play simultaneously throughout the whole of Physics. But the question still remains as to the exact significance of this dualism of waves and corpuscles. The problem is extremely difficult, and is far from being fully resolved.

The simplest view is that advanced by Schrödinger at the beginning of his researches, according to which the corpuscle, or electron, consists of a wave-group, or *Wellenpaket*. We saw that this was a tenable view so long as we confine ourselves to those mechanical phenomena which are in accord with the older dynamics, i.e., in terms of the new Theory, to phenomena where the propagation of the associated wave obeys the laws of geometrical optics. Unfortunately, when we pass to the sphere of the new Theory itself, this idea, attractive as it is by its very simplicity, appears incapable of being sustained. In an experiment like that of the diffraction of an electron by a crystal the wave-group would be completely dispersed and destroyed, so that we should not be able to find any corpuscles in the various scattered beams. In other words, if they were simple wave-groups, corpuscles would have no stable existence.

It seems impossible then to maintain Schrödinger's idea in its entirety. Another notion to which the author was attracted for a long time, and according to which the corpuscle is a singularity in a wave phenomenon, seems no easier to work out. In the special case where the corpuscle is in uniform motion, it is possible to find a solution of the equation for waves which represent a singularity in motion and can be employed to represent the corpuscle. But it is difficult to generalize this in the case of motion that is not uniform; and there are serious objections to this point of view.

Another attempt made by the author is recorded in his Report to the Fifth Solvay Congress; this is based on the suggestion that, since we must always associate a wave with a corpuscle, the idea which is in the closest agreement with the older physical concepts is that the wave is a real and actual phenomenon, occupying a certain region in space, and that the corpuscle is a material point having a certain position in the wave. I have stated already that the intensity of the wave at any point must be proportional to the probability that the corpuscle is present there; and accordingly we must try to connect the motion of the corpuscle with the propagation of the wave in such a way that this relation is realized universally and automatically.

Actually it is possible to establish a connection between the motion of the corpuscle and the propagation of the wave; for provided that at the initial instant the intensity of the wave measures at every point the probability that the corpuscle is present there, then it also indicates this probability at every later instant. The corpuscle may thus be regarded as guided by the wave—a kind of pilot-wave. This view provides an interesting picture of the motion of corpuscles in Wave Mechanics without there being any need to abandon classical ideas too sweepingly. Unfortunately, however, here too very serious objections are met, and it is impossible to remain satisfied with the concept of the wave as a kind of pilot. At the same time the equations on which this theory rests cannot be challenged, so that some of its results can be preserved by giving them a less uncompromising form in accordance with ideas independently elaborated by Kennard.[1] Instead of speaking of the motion and the trajectory of corpuscles we speak of the motion and trajectory of the "elements of probability," and in this way the difficulties mentioned are avoided.

There is, finally, a fourth point of view, which is the one most in favour at present, developed by Heisenberg and Bohr. It is certainly slightly startling at first sight: it seems, none the less, to contain a considerable element of truth. On this view the wave is in no sense a physical phenomenon in any region of space; it is rather a mere symbolical representation of what we know about the corpuscle. Experiment or observation, therefore, can never enable us to say precisely that a given corpuscle occupies such and such a position in space, or that its velocity has such and such a magnitude and direction. All that experiment can teach us is that the position and velocity of the corpuscle lie between certain limits: in other words, that there is a calculable probability that it has a certain position, and also a probability that it has a certain velocity. The information given us by an initial experiment or observation, made at a certain instant t_0, can then be represented symbolically by a wave, the intensity of which at this instant t_0 gives us, at every point, the probability that the corpuscle is present at that point, while its spectral structure gives us the relative probability for the various states of motion. If we investigate the propagation of the wave from the instant t_0 to a later instant t, then the distribution of the intensities and the spectral structure of the wave, at the instant t, will enable us to say what is the probability that a second observation or experiment, made at the instant t, will localize the corpuscle at a certain point, or attribute to it a certain state of motion.

The fundamental consequence following on this treatment of the problem is Heisenberg's Principle of Uncertainty. A specific wave-train can be regarded as monochromatic only if its dimensions are large relatively to the wave-length. Accordingly if a corpuscle

[1] *Physical Review*, XXXI, 1928, p. 876.

is localized by observation in a region whose dimensions are not large relatively to the wave-length, then we shall have to represent the corpuscle by a wave-train which will be far from being monochromatic. It follows that, from Heisenberg's point of view, the more exactly we wish to determine the position, the less exactly will the motion be determined. Conversely, the more exactly the motion of the corpuscle is defined, the more closely will the associated wave approximate to a plane monochromatic wave of constant amplitude. Hence the more exactly the state of motion is defined, the less exactly will it be possible to define the position of the corpuscle.

Bohr maintains, still further, that there are "two complementary aspects of Reality," localization in space-time, and dynamic specification in terms of energy and momentum. These are like two distinct planes which we cannot exactly focus simultaneously. To take an illustration, let us imagine a pattern parts of which are drawn in a plane Π while other parts are in a plane Π' parallel and very close to the former. If we look at the pattern through an optical instrument of moderate precision, we shall obtain a tolerably clear outline of the pattern by focussing on a plane lying between Π and Π', in which case we shall have the impression that the whole of the pattern is drawn in a single plane. But if, on the other hand, we use a very exact optical instrument we shall not be able to focus on Π and Π' simultaneously; the more exactly we focus on Π the worse will be the image we obtain of the parts of the pattern drawn on plane Π', and conversely; so that we shall be compelled to recognize that the pattern is not drawn on one plane. Thus the older Mechanics could be compared to the moderately precise instrument, and it gave us the illusion that we could define exactly and simultaneously the position and the state of motion of the corpuscle. With the new Mechanics, however—which is to be compared to the high-precision instrument—we are forced to recognize that the localization in space and time on the one hand, and the specification in terms of energy on the other, are two different planes of Reality which we cannot seize upon simultaneously.

Such, it would seem, is the fundamental idea of Bohr and Heisenberg; and this point of view implies a consequence which had been predicted by Born, viz. that we can no longer affirm the existence of a strict Determinism in Nature, since the entire Determinism of the older dynamics rested on the fact that it was possible simultaneously to determine the position and the initial velocity of a corpuscle, which according to Heisenberg's Principle is impossible. Accordingly there are no strict laws: all that we have are laws of probability.

A number of curious circumstances are implied in this interpretation of Wave Mechanics. In the first place, the corpuscles exist, and it is still assumed that there is a certain significance in discussing them; at the same time, if we adopt Bohr's views, we are no longer able to regard them in the simple classical way which consists in treating them as minute objects having a position in space, a velocity and a trajectory. Secondly, the other term of the duality, the wave, is now simply a purely symbolic and analytical representation of certain probabilities, and is not a physical phenomenon in the old sense of the term. The last point can be clearly shown by an example. Let us assume that at the instant t the wave-train associated with a corpuscle occupies a region R, and that a certain observation made at that instant enables us to assert that the corpuscle is in a region R', which of course is included within R. In that case the wave packet ought to be "reduced," according to Heisenberg's expression, i.e. all that part of the wave which is within R, but outside R', vanishes as vanishes the expectation of an event which is not realized. This shows clearly the non-physical character of the wave according to Bohr's and Heisenberg's conceptions.

To sum up, the physical interpretation of the new Mechanics remains an extremely difficult subject. At the same time one important fact has now been thoroughly established, viz. that both for Matter and for radiation we must assume the dualism of waves and of corpuscles, and that the spatial distribution of corpuscles can only be predicted by taking into consideration the concept of waves. Unfortunately, the ultimate nature of the two terms of the duality and the exact relation subsisting between them are still extremely obscure.[1]

[1] Since this was written, the interpretation of the New Mechanics in terms of Probability by Bohr and Heisenberg has been practically universally adopted by theoretical physicists, and is more fully discussed in Part V.

3

THE PASSAGE OF ELECTRICALLY CHARGED CORPUSCLES THROUGH POTENTIAL BARRIERS [1]

THIS Chapter is concerned with the passage of electrically charged corpuscles through potential barriers, and the theories connected with this phenomenon. It is a subject that provides a good illustration of the methods of Wave Mechanics, and it shows how thoroughly the problems of the dynamics of electrons have been transformed into problems closely resembling those of classical optics.

To bring out the special interest of the question, we must begin by recalling how the notions of the corpuscle and the wave came to be amalgamated in Wave Mechanics. Quite recently it was thought that the corpuscles forming Matter—electrons and protons —might be considered as being just electrified material points, obeying the laws of classical dynamics, which were corrected when necessary by the principles of the Theory of Relativity. In the same way, the more complex structures known as atoms and molecules could also be treated as material points, following the laws of the older dynamics, whenever their motion *en masse* had to be dealt with. The study of Matter would thus consist essentially in tracing the motion of elementary corpuscles with the aid of the laws of classical dynamics.

The method to be used for investigating luminous phenomena, however, was wholly different. For here there were no corpuscles in a state of motion, the variation of whose co-ordinates the laws of dynamics enabled us to deal with as a function of the time;

[1] Opening Lecture of a course delivered at the Institut Henri Poincaré during the academic year, 1931–32.

instead, there were waves propagated in accordance with a partial differential equation of the type $\Delta u = \dfrac{1}{V^2}\dfrac{\partial^2 u}{\partial t^2}$, which is the generalized equation of vibrating cords. The velocity of propagation, V, is characteristic of the medium of propagation of the wave. In homogeneous media—a vacuum for example—the velocity is everywhere the same, though it may vary from one point to another in heterogeneous media. This velocity V is defined by the index of refraction, which gives at every point the relation between the velocity in a homogeneous medium of reference (generally a vacuum) and the velocity V.

We have repeatedly found that the nature of a light-wave has been imagined in a good many different ways; Fresnel and those who followed him regarded it as the vibration of the ether, an elastic and exceedingly subtle medium which was supposed to penetrate all bodies. Clearly the properties of this medium would have to be rather strange, since it was supposed to offer no resistance to the motion of bodies and yet was supposed to be more rigid than steel. Later, with Maxwell, luminous vibrations were believed to be of an electromagnetic character. The interpretation of the light-wave which treated the latter as the mechanical vibration of a certain medium was embarrassed by this view, and it was abandoned altogether with the development of the Theory of Relativity, which requires all universal media of reference, like the ether, to be discarded. At the same time the Wave Theory of Light was based essentially on the equation governing the propagation of waves, and had been verified with the utmost exactness by experimental research into the phenomena of interference and diffraction: and it retained its value to the full. The essential point of the Theory is wholly independent of any elastic, electromagnetic or other interpretation given to it, and it consists in the principle that the energy localized at a given point of the light-wave is proportional to the square of the amplitude of the wave at that point: that is to say, to the intensity of the wave. We have thus two classes of phenomena. We have the motion of material particles on the one hand, and the propagation of

Light on the other: and the two appear to obey essentially different laws. We study the motion of particles by tracing the variation of their co-ordinates with time, this being determined by the differential equations of Mechanics.

For any given particle, its motion is completely represented by only a single curve—its trajectory. Further—apart from the quite specific group of periodic motions—there is no time interval and no length capable of playing any particular part in the journey of the particle. There is nothing in the classical view to make us suspect that the motion of particles has any analogy with the phenomena of diffraction, interference and resonance which are so characteristic of the Wave Theory, and which introduce integers into the formulae.

In the classical Wave Theory, on the other hand, propagation involves an extended spatial region, and not merely a line. It is governed by a partial differential equation for the very reason that, in any given region, the wave is represented by a function $u(x, y, z, t)$, while for the material particle the old ideas give us three functions of time, $x(t)$, $y(t)$ and $z(t)$, which vary with the motion of the entity. Further, we know that any wave can be decomposed into a superposition of simple sine waves having well defined frequencies. In optics each frequency corresponds to one simple colour, which is why the sine waves are known as mono-chromatic waves, and apparatus (prisms and gratings) exists effecting this decomposition into simple colours, which is always mathematically possible by Fourier's Theorem. A monochromatic wave in a homogeneous medium is defined by its frequency ν and its wave-length λ, the latter following from the former in accordance with the equation $\lambda = \dfrac{V}{\nu}$, where V is the velocity of propagation in a homogeneous medium. In certain cases, the propagation of the monochromatic wave may be powerfully affected by the wave-length. For example, if in a homogeneous medium certain obstacles are placed in the path of the wave, the influence exerted by these obstacles on the propagation will be entirely different according as certain dimensions of the obstacles are, or are not, integral multiples of the wave-length. Here then we have a number of essential differences between the motion of a corpuscle and the propagation of a wave. But if we examine them carefully, these differences are perhaps less profound than a superficial survey might lead one to believe. We must notice first that the motion of a corpuscle depends on the field of force within which it is placed, i.e. to some extent on some state of the medium within which its journey proceeds. In Physics this state is most frequently defined by the potential function of which the force is the gradient, in which case, in the Newtonian equations, the second terms contain the partial derivatives of this function. It is therefore misleading to say, as I have just done, that the motion of a corpuscle involves only a line—a spatial curve; for the fact is that this line—the trajectory—is determined by the character of the field of force in the neighbouring region. This is clearly perceived if we examine Maupertuis's Principle of Least Action: in a steady field of force derived from a potential function $U(x, y, z)$, the trajectory of a corpuscle of energy E passing through two points A and B is such that the integral

$$\int \sqrt{2m(E - U)}\, ds$$ along the trajectory from A to B is stationary;

and, to apply this Principle, we must take into consideration the curves passing from A to B, which are infinitely close to the trajectory. This causes the properties of the field in the entire region adjacent to the trajectory to be involved.

It was especially Hamilton's researches which, about a century ago, illuminated the profound analogies which under certain conditions exist between the motion of corpuscles, considered according to Newtonian dynamics, and wave propagation. We know that if the index of refraction does not vary too rapidly from point to point in the medium, the propagation of a wave can be traced by those methods of approximation which in fact constitute geometrical optics. We define the rays of the wave propagation by Fermat's Principle of Minimum Time,[1] and the totality of the rays forms a totality of curves orthogonal to a family of surfaces known

---

[1] Or, once again, Stationary Time; cf. p. 123.

as wave-surfaces. Although the word "wave" is used in the latter description, geometrical optics is entirely independent of any undulatory concept, and the propagation of Light is studied by it in accordance with purely geometrical methods and principles. The rays are trajectories of energy, and there is nothing to prevent us from treating the totality of light-rays as being the totality of the trajectories described by minute projectiles which would constitute Light in accordance with Newton's views. At the same time geometrical optics is valid only for free propagation (without obstacles) in a medium having a refractive index gradually variable on the scale of the wave-length. If, however, the index is changed very suddenly, as at the surface of separation of two media, or if there are obstacles in the path of the Light such as screens, then geometrical optics becomes unable to explain the phenomena that arise and, as has been known since the days of Fresnel, it is the Wave Theory alone which can interpret them (interference, diffraction, thin films). Fortunately for the consistency of our physical explanations, however, geometrical optics proves in this investigation to be a method of approximation whose use is justified by the wave theory whenever propagation is free and the index gradually variable. None the less the concept of a ray becomes meaningless whenever there is interference or diffraction, and accordingly since Fresnel's day the corpuscular Theory, which so far had appeared to be compatible with geometrical optics, has had to be abandoned.

The great service rendered by Hamilton was to have perceived the complete formal analogy that subsists between geometrical optics and Newtonian corpuscular dynamics. To the rays determined by the Principle of Minimum Time there correspond the trajectories determined by the Principle of Least Action, while to the totality of rays propagated together, there corresponds the totality of trajectories which analytical Mechanics associated with one and the same Hamilton-Jacobi function; to the wave-surfaces correspond the surfaces S = the constant which is the value of the Jacobi function. At the same time some important points remain where the analogy does not appear to hold good. This is true particularly of the part played by the velocity of the corpuscle in dynamics, and by the velocity of the wave in geometrical optics. These were the difficulties which prevented Hamilton seeing in the analogy he had discovered anything more than a merely formal analogy. Had he realized that the analogy was more than simply formal and had some physical significance, to what conclusions would he have been led?

Newtonian Mechanics, then, corresponds to geometrical optics, while the latter in turn is only an approximation to undulatory optics. Accordingly Hamilton ought to have concluded that Newtonian Mechanics had only a limited validity, that the true Mechanics ought to have an undulatory character, and that Newtonian Mechanics can be accepted only as a first approximation valid in certain cases, just as the Wave Theory of Light assumes that geometrical optics is a first approximation valid in certain cases. This bold idea, however, which found no place in Science until a century later, implied one grave difficulty. Exactly as the concept of a ray vanishes when we pass from geometrical optics to wave optics, so also the concept of a trajectory ought to vanish as we pass from Newtonian dynamics to Wave Mechanics, and consequently, in the latter system, the concept of the corpuscle ought, if not to vanish, at any rate to undergo a strange transformation.

We are already familiar with the discoveries which led to the development of Wave Mechanics. The series originated with the Quantum Theory which, in turn, was based on Planck's researches on black body radiation. The essential idea of this Theory from Planck to Bohr, Sommerfeld, etc. was that the motion of particles at any rate on a very small scale, cannot be predicted completely by the sole use of Newtonian dynamics. Of the entire motion whose possibility can be predicted by Newtonian dynamics, there is only a restricted group which actually exists in Nature, and these privileged or quantized movements are defined by the fact that certain magnitudes which characterize them, having the dimensions of action (energy $\times$ time, ML^2T^{-1}), are integral multiples of Planck's quantum of action, h. This entry on the scene of integers, while wholly natural in the various Wave Theories, appears inexplicable in dynamics; but it has since been realized that it indicates the need for substituting a new Wave Mechanics for the older Newtonian geometrical dynamics, just as in the Theory of Light, wave optics has had to be substituted for geometrical optics. The difficulty now, however, will be that of preserving the concept of the corpuscle, since we can no longer exactly define its trajectory.

Another class of new facts has helped to give a precise direction to the evolution of our concepts. I have in mind phenomena like the photo-electric effect and Compton effect, which have demonstrated the existence of corpuscles of luminous energy equivalent to $h\nu$ in all radiation of frequency ν; or to put it more cautiously, these phenomena have shown that in certain cases things happen exactly as though radiation of frequency ν consisted of corpuscles of energy $h\nu$. All this agrees very well with the Newtonian Emission Theory, and we are forced to revert to the idea that such things as corpuscles of Light exist, despite the undeniable existence of diffraction and interference. Here again corpuscular notions have to be preserved to some extent, despite the fact that in the general case (i.e. outside the domain where geometrical optics applies) we cannot ascertain the trajectories of the corpuscles. The difficult compromise between the concept of corpuscles and that of waves, as simultaneously valid for Matter as well as for Light, has been effected by Wave Mechanics—but only at the cost of accepting entirely novel ideas and of a certain renunciation of the clear concepts and complete Determinism of classical Theory. Heisenberg's Uncertainty Principle has shown us to what a degree this renunciation must be accepted if we wish to reconcile the existence of corpuscles with a wave dynamics which, in the general case (i.e. apart from the conditions corresponding to geometrical optics) cannot exactly define the trajectory of corpuscles nor their motion.

But what is particularly interesting at the moment is the fact that the new Mechanics, by submitting the dynamics of the corpuscle to undulatory demands, has enabled us to predict extremely interesting phenomena which Newtonian dynamics failed to explain.

First 'among these is quantized small-scale motion, e.g. the quantized motion of electrons within atoms. Wave Mechanics has achieved a complete interpretation of these, by which they are treated as analogous with the phenomena of resonance or of stationary waves; the appearance of integers in the specification of these quantized states then appears entirely natural. Another phenomenon whose discovery has provided direct experimental proof in support of the guiding idea of Wave Mechanics is the diffraction of electrons by crystals, which is entirely similar to the Laue-Bragg phenomenon for X-rays.

But the new Mechanics also enables us to predict other types of phenomena whose theoretical study is extremely interesting because of their analogy with the classical optical phenomena. Take for example one fundamental optical phenomenon, that of thin films. Let us place, normal to the trajectory of a plane light-wave of frequency v, a homogeneous film with parallel surfaces of a refracting substance, and let λ be the length of the light-wave *within the film*. Then if the thickness of the film is equal to an odd multiple of $\frac{\lambda}{4}$, the waves reflected at the film surfaces of entrance and exit will be in the same phase, and there will be intense reflection. If on the other hand the thickness of the film is an even multiple of $\frac{\lambda}{4}$, i.e. if it is an integral multiple of $\frac{\lambda}{2}$, then there will be feeble, or no, reflection. A familiar method of producing these phenomena is to use a wedge-shaped film, i.e. one of variable thickness. In such a case we shall obtain bright fringes and dark fringes, corresponding to the places where there is intense reflection or almost complete transmission, according to the thickness of the film; Newton's rings are an instance of this. It is interesting to recall once again the way in which Newton, who favoured the corpuscular theory, tried to explain these rings of his discovery. He assumed that the light-corpuscles, once they had entered the refracting medium, passed alternatively and periodically through "fits of easy reflection and easy transmission." It could thus be seen that, according to the thickness of the film, the corpuscle on reaching the second surface of separation could be in either a state of easy reflection or else of easy transmission. The Wave Theory has, however, provided a simple explanation of these film phenomena, and has caused Newton's corpuscles and their varying "fits of easy reflection and transmission" to be completely forgotten.

In the case of material corpuscles, too, we can observe phenomena entirely analogous to those observed with thin films. In the new Mechanics what corresponds to a homogeneous refracting medium is a region of space where the potential is constant. I shall discuss these cases, and also the more general ones where the potential is not constant, which correspond to the passage of Light through heterogenous refracting media. Connected with these theories, again, are the attempts to explain the Ramsauer effect and the abnormal scattering of X-rays.

Another class of phenomena predicted by Wave Mechanics within the same field of research is connected with the fact that in the Wave Theory there is no barrier which a wave cannot cross. The first example of this is found in optics in the Theory of Caustics. In geometrical optics the caustic is the curve or envelope enclosing the light-rays, and constitutes an absolute limit for the field of luminous energy. Now if the problem of caustics in undulatory optics is closely examined we find that, roughly speaking, the caustic is the limit of the luminous energy, but that in absolute strictness there is a minute penetration by the waves, and consequently by the luminous energy, into the region outside the caustic. In Wave Mechanics what all this signifies is this. Let us take a region in which a uniform field prevails. At one point in this field there are projected corpuscles under the influence of the field, all having the same initial velocity, and moving in every direction. The problem is to ascertain their motion. In the older Mechanics this problem is that of the motion of projectiles within a gravitational field (air resistance of course being neglected), and their trajectories are the parabolas whose axis is parallel to the direction of the field. These parabolas can be enveloped within another parabola, the "safety parabola," outside of which it is impossible to be struck by a projectile—hence the name (Fig. 1). The safety parabola is a caustic, and it is also a limit to the energy of the salvo of projectiles —a limit which cannot be passed. Now if the same problem is treated in Wave Mechanics, as applied to corpuscles, we shall find that the chances are certainly very slight of finding one of these corpuscles outside the safety parabola. But at the same time there will be a *slight* chance that a corpuscle will be encountered in the region outside the caustic parabola. In the same way the geometrical shadow along the edges of a screen does not constitute an absolute limit for the light, since the theory of diffraction by the edge of a

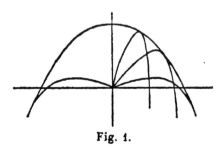

Fig. 1.

screen shows that there is some light within the geometrical shadow. In the same way, again, when total reflection of Light occurs there is a "damped" wave which penetrates slightly into the second medium. In all these cases, then, there is no absolute limit for these waves, which always decrease *continuously*, however rapid that process may frequently be. Analogous instances are found in Wave Mechanics as applied to corpuscles; the waves associated with corpuscles will never be completely stopped by an obstacle; and since a corpuscle can be found at any point where its wave is not equal to zero, no corpuscles can ever be completely stopped by any barrier. Here again we have a fundamental difference as compared with the older Mechanics. For let us take a corpuscle with energy E, which approaches a region where the potential is increasing, reaches a maximum and then decreases. (Fig. 2).

We have here a mountain of potential. While ascending this mountain the corpuscle will lose its kinetic energy, and if the value of the potential at the summit of the mountain is greater than E the corpuscle will stop, since its kinetic energy cannot become negative, and will then return without having been able to cross the mountain. Thus in classical Mechanics the mountain of potential is a barrier which corpuscles of slight energy can never cross. But Wave

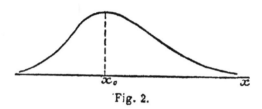

Fig. 2.

Mechanics envisages a wholly different phenomenon. The wave associated with the incident corpuscle will be reflected almost completely by the mountain of potential, but none the less a minute fraction of the wave will penetrate to the region on the right in the diagram. The interpretation of this fact is as follows. The preponderant chances are that a corpuscle will be reflected by the mountain of potential and thrown back to the left; but however high the mountain may be, there is always a minute chance that a corpuscle may pass to the right. In other words, given a cloud of corpuscles of energy E approaching the mountain of potential from the left, practically all of them will be reflected to the left,

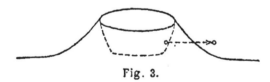

Fig. 3.

but there will always be a few which will succeed in crossing the mountain. It follows that if there are corpuscles in a closed region of space, and if there is a field at the limits of this region to prevent them from leaving it, then there will nevertheless always be some which will succeed in escaping, however high the excluding walls of potential may be.

When there are corpuscles enclosed within a kind of circus

surrounded by a mountain of potential, some will always have a chance to get into the open, however high the mountain may be. (Fig. 3). Such is the principle of the Theory of Radioactivity advanced by Gurney and Condon and by Gamow, a Theory with which various other tentative theories are linked up. It is also the principle of a Theory of Catalysis formulated by Born.

4

RELATIVITY AND QUANTA

THE development of contemporary theoretical Physics has been characterized by the birth of two important theories, the Relativity and Quantum Theories. To a great extent they have developed independently; but each of them is comprehensive and claims validity for the whole of Physics, so that a mutual encounter and confrontation was inevitable. Here I propose briefly to recall their independent earlier development, and then to show what attempts have been made to effect a compromise between them— attempts which so far have not met with complete success.

*   *   *

The two theories owe their origin to the fact that certain discords were found to exist between earlier theories and experimentally established facts. The discord which led to the Theory of Relativity is familiar—the impossibility of demonstrating the absolute motion of the Earth relatively to the ether. Classical Mechanics maintains that the observation of purely mechanical phenomena does not suffice to allow an observer to determine whether he is at rest or in a state of uniform movement in a straight line with reference to the totality of the fixed stars. But until recently it was thought that if we leave purely mechanical phenomena and pass to optical phenomena the case would be altered. Since the time of Fresnel it had in fact been assumed that Light was a perturbation akin to that which is the cause of sound, and it seemed inconceivable to the physicists of that day that luminous perturbations propagated through space should do so without the intervention of a certain medium. But since this medium escapes our senses, and since it must exist in space and not interfere with the motion of the stars, physicists were compelled to imagine it as infinitely subtle, as ether, and differing absolutely from the material media perceived by our senses. Later, as the theory of electromagnetic phenomena developed and, thanks to Maxwell, was extended in the manner previously explained, physicists came to believe that the electric and magnetic fields propagated in space also have a supporting medium, and that, since Light is no more than an electromagnetic wave, electromagnetic phenomena, too, must have the ether as medium. Unfortunately the concept of the ether met with all sorts of difficulties, and all the many attempts made to envisage an elastic ether or to imagine its exact structure led to rather unsatisfactory conclusions. Gradually physicists gave up any attempt to define its nature exactly and confined themselves to treating it as a mere medium of reference, a medium supporting the electromagnetic phenomena. Such was the point of view most frequently assumed by Lorentz in elaborating his Theory of Electrons. Unfortunately the concept of the ether has proved fairly embarrassing even when reduced to this minor part. The position is that if there is such a thing as a medium of reference or support for electromagnetic and optical phenomena, then these phenomena cannot manifest themselves in the same way both for an observer at rest in this medium, and for another in motion relatively to it. Nor can a physicist in his laboratory, by investigating purely mechanical phenomena, detect the motion of the Earth in its orbit; but if the ether exists, he ought to be able to do so by the study of certain optical or electromagnetic phenomena. Now experimental attempts have obstinately refused to reveal the motion of the Earth relatively to the ether. The famous experiment made by Michelson, and others of the same kind, which according to the older theories ought to have given a positive result, produced in fact a negative one. A number of somewhat artificial explanations were at first advanced (e.g. FitzGerald's contraction): then came Einstein, who cut the Gordian knot. He assumed that for observers in a state relatively to each other of uniform motion, all the phenomena of Nature—optical, electromagnetic and mechanical—obey the same Laws, with the result that none of the observers can, by means of observations made within his private system, demonstrate his own movement, and that each of them has an equal right to consider himself at rest. From this Principle of Relativity, and from a piercing analysis of the methods used for spatial and temporal measurements, Einstein deduced that the Time and Space co-ordinates used by each individual observer are interconnected by the formulae of transformation known under the name of Lorentz Transformations. These relations between the co-ordinates used by observers in a state of relatively uniform motion have been systematized by Minkowski, who envisaged a continuum of four dimensions—the Universe, or Space-Time. Each observer carves *his own* private space and time out of this continuum in a certain way, and this carving is not executed in the same way by any two observers in a state of motion relatively to each other.

The development of the Theory of Relativity has led to a kind of close fusion between Time and Space, which is symbolized by Minkowski's Space-Time; and in all the relativistic formulae the Time and Space co-ordinates play a very symmetrical part. Nevertheless it would be an exaggeration to say that Time and Space play wholly symmetrical parts in the Theory of Relativity. For one thing a difference between them appears in the formal statement of the Theory, since it is the quantity $\sqrt{-1}\, ct$ which there plays a part symmetrical to that of the variables x, y, z; and for another, of course, no theory can eliminate the two fundamental facts that, while the Space variables can vary in one sense or the other indifferently, Time always flows in one sense only, and that elementary physical entities, like atoms and electrons, etc., persist through Time, which fact is expressed on the macroscopic scale by the persistence of material objects, with the result that Space-Time has a kind of "linear" structure in the direction of Time.

Among the important consequences of Relativity, still further, there is the concept of the inertia of energy, according to which a mass $\dfrac{W}{c^2}$ is associated with every quantity of energy W, whence it follows that if it is in a state of motion with velocity v within the system of reference employed, a momentum $\dfrac{W}{c^2} v$ is associated too. The Principle of Relativity also involves an important change in the old dynamics of the material point. Relativistic Mechanics diverges widely from classical Mechanics as far as corpuscles having velocities approximating to that of Light in empty space (c) are concerned. Incidentally it seems to follow that c is the upper limit of the velocities obtainable by a material body, since according to relativistic Mechanics such a body would have to be supplied with infinite energy in order to attain velocity c. The formulae of relativistic Mechanics for electrons of high velocity have been supported by a series of experiments, of which the most decisive were those made by Guye and Lavanchy.

In its earlier form, the Theory of Relativity was concerned with the Space and Time co-ordinates only of observers in a state of uniform motion in a straight line. Later it was generalized by Einstein himself, and has provided a description of Gravitation. This generalized Relativity, however, I shall here leave on one side.

*   *   *

Experimental methods, as we saw, obstinately refused to reveal the Earth's motion relatively to the ether, thus compelling physicists to accept the Principle of Relativity; almost at the same time, too, experimental methods caused them to introduce into their previous views a modification which certainly is even more important. I have in mind the idea of the quantum.

The Quantum Theory originated in research on the distribution of energy between the frequencies in radiation in thermodynamic equilibrium. The classical theories envisaged a certain law of spectral distribution (Rayleigh-Jeans law), and this law, which leads to a startling consequence in giving an infinite value to the total density of energy, has turned out to be correct only for the lower frequencies. Now experiments provided curves of the spectral distribution of energy in disagreement with theory, but at the same time very clearly determined, and consequently there was a justification for attempts to revise theory so as to obtain the result already given by experiment. This was the work of Planck, who succeeded in discovering the law of spectral distribution which bears his name, and which accounts perfectly for the facts experimentally observed. To reach this success, and more particularly to introduce his famous constant h, of which the classical theories were unaware, Planck had to found his arguments on the novel hypothesis of the existence of quanta. Originally, as we have already found, he envisaged the hypothesis in the following form. He assumed that the periodic motion of electrified elementary particles is stable only if the energy E of their oscillations is connected with the frequency v by the equation $E = hv$. In this form we have a hypothesis of quanta of energy, and it was with its assistance that Planck deduced his Law of black body radiation. The hypothesis of energy quanta, however, can be applied only to mechanical systems whose frequency of oscillation is independent of the energy, which was precisely the case of the elementary oscillators first considered by Planck. In searching for a more generalized formulation of the Quantum Theory, Planck soon realized the necessity of presenting it in the form of a theory of quanta of action. In this way the idea was reached that the constant h revealed the existence of a kind of atomism for mechanical action.

As I have pointed out previously, the magnitude called "action" in Mechanics has the dimensions of energy multiplied by time or, what amounts to the same thing, of momentum multiplied by length. Action is thus a magnitude depending *simultaneously* on the configuration of the system and on its dynamic state. Hence follows the fundamental fact—of which Heisenberg's Uncertainty Principle in the new Mechanics is ultimately but one aspect—that the existence of the quantum of action expresses the fact that it is impossible to consider the configuration of the system and its state of motion separately and independently of each other. The resulting link between geometry and dynamics is disconcerting to our minds, and is wholly alien to classical views; nor does it fit in with the relativistic representation of motion by curves (World-lines) in Space-Time. Such facts might have created the impression that in the atomic sphere, where the influence of the quantum of action is predominant, there would be some difficulty in reconciling Relativity and quanta.

An examination of a large number of physical phenomena has shown the importance, quite apart from black body radiation, of the quantum of action, whose existence is manifested by the intervention of the constant h. More particularly, Bohr's famous atomic Theory, on whose successes it is superfluous to dwell, is founded on the quantizing of periodic motion. The form in which this quantizing was obtained was due to the introduction, in the best practicable way, of the idea of the quantum into the equations of the older Mechanics.

*   .   *
*

The original development of the Quantum Theory (between 1900 and 1916) took place quite independently of Relativity Principles. None the less there is one aspect of the Quantum Theory—an aspect pointed out by Einstein—which has some relation with Relativity. We have seen that in 1905, at the very moment when at the age of 25 Einstein was laying the foundations of the Theory of Relativity, that scientist, by a brilliant intuition, had also sensed that the existence of certain phenomena of interaction between Light and Matter (the photo-electric effects) implied the corpuscular structure of light-energy. These phenomena have already been discussed in some detail;[1] and whenever we are dealing with Light, i.e. with a form of energy propagated in a vacuum with velocity c, we may expect Relativity to come into play. This intervention manifests itself in the following manner. If we assume, once again, that Light of frequency ν is divided into corpuscles of energy $h\nu$, we must attribute to each of these a momentum $\dfrac{h\nu}{c}$, since relativistic Mechanics requires that the energy W and the momentum p of a corpuscle with velocity v are connected by the equation $p = \dfrac{W}{c^2}\, v$. For light-corpuscles with

[1] P. 180.

velocity c this gives us $p = \dfrac{W}{c} = \dfrac{h\nu}{c}$; and in this way the new corpuscular Theory of Light (the Theory of light-quanta, or photons) has, so to speak, provided the first meeting place for Relativity and quanta, the quantum of action finding its expression in the value $h\nu$ assigned to the energy of the light-corpuscle, and Relativity coming into play by giving the value of the ratio between momentum and energy, and thus imposing on the momentum the value $\dfrac{h\nu}{c}$. The older Mechanics, on the other hand, would have given a completely different value.

The Relativity and the Quantum Theories met on another occasion in 1916, when Sommerfeld formulated his theory of the fine structure of the Hydrogen spectrum and of the regular doublets of X-ray spectra. Bohr had already succeeded in giving a rough explanation of the Hydrogen spectrum, of ionized Helium and even of the X-ray spectra, by an application of Newtonian Mechanics (to which the postulate of the quantum of action had been added) to the planetary model of the atom. These various spectra, however, manifest a fine structure which Bohr's Theory failed to explain. It then occurred to Sommerfeld to apply to atomic electrons the relativistic Mechanics of Einstein instead of the classical Mechanics of Newton, retaining at the same time the postulate of the quantum of action. In this way he was enabled to predict correctly the fine structure of the spectra of Hydrogen and of ionized Helium, as well as certain important details of X-ray spectra (the regular doublets). All this looked like a notable success, which was due to the partnership, to put it that way, between Relativity and quanta. Unfortunately, on examination, Sommerfeld's theory proved to be less satisfactory than had at first been thought, for while it certainly finds a place for the fine structure of X-rays, this is not the place found in reality; Dirac's Theory of the magnetic electron was required to set things right.

* * *

We have seen repeatedly how important a turning point it was in the history of contemporary Physics when the new Mechanics appeared on the scene. Its entry took place nearly simultaneously in two different forms—in that of Quantum Mechanics, and of Wave Mechanics. But whereas Quantum Mechanics at once revealed a definitely non-relativistic character, Wave Mechanics (on the other hand) owes its origin to considerations based on Relativity, since the parallelism between the motion of a corpuscle and the propagation of a wave, on which Wave Mechanics is founded, had been elaborated in accordance with Lorentz's transformation formulae. At this point, however, a curious thing happened. When the attempt was made to work out the new undulatory concepts of Mechanics mathematically, it became necessary in a measure to abandon the relativistic domain, and to formulate a Wave Mechanics which is in fact a development of Newton's, and not of Einstein's, Mechanics. A most curious phenomenon, indeed, since it means that Wave Mechanics appeared to be denying its own origins!

When developed under its non-relativistic form the new Mechanics, as we know, proved very successful. Its triumphs need not be enumerated here. It led to an interpretation of the Laws of Physics in terms of probability, an interpretation which after some initial resistance seems now to be accepted fairly generally. This probability interpretation is expressed in a brilliant generalized theory which enables us to predict the possible values of a physical magnitude and their respective probabilities.

But despite the undoubted success of the new Mechanics, the fact that it was essentially non-relativistic remained sufficiently disconcerting. Ever since the beginning of the full mathematical development of Wave Mechanics a certain number of authors have attempted to give the generalized equations of the new science a relativistic form. It was, however, not a happy attempt. On the one hand the relativistic equations in question lead to formulae different from those of Sommerfeld, and in contradiction with the results of experiment so far as the fine structure of the Hydrogen and the X-ray spectra is concerned; on the other hand, the very form of the suggested equations made it impossible to conserve the probability theory in its generalized form already referred to; and this constituted a serious objection. In consequence of this failure of the first attempts to give a relativistic form to Wave Mechanics, it seemed for a moment as though the new Mechanics would take no further interest in Relativity. Soon, however, a new and striking effort was made by Dirac (in 1928) to reconcile Quantum Mechanics and Relativity, in his construction of the brilliant Theory of the magnetic moment and spin of the electron, with which I shall now deal in fuller detail than in my previous discussions.

* * *

Dirac's aim in elaborating his Theory was to remain in harmony with Relativity principles, and at the same time with the generalized probability interpretation of the new Mechanics. Without dwelling unduly on the mathematical aspect of the Theory, I may yet point out that Dirac reached his objective in an extremely ingenious manner by "linearizing" the classical Hamiltonian function, with the result that the foundation of the wave dynamics of the electron ceased to be a partial differential equation of the second order for a single wave-function, its place being taken by four simultaneous partial differential equations of the first order for four wave-functions. The most remarkable feature of this Theory is that it automatically introduces the magnetic moment and angular momentum (spin) of the electron—two new elements whose consideration had already previously been rendered imperative by experimental results. Thus Dirac's Theory gives correctly the fine structure of Hydrogen and X-ray spectra without raising the difficulties inherent in Sommerfeld's earlier formula, and also yields a correct interpretation of the abnormal Zeeman effects, which none of the previous theoretical attempts had succeeded in obtaining.

But while Dirac's Theory has met with a great measure of success, it would be a mistake to think that it has effected a complete reconciliation between quanta and Relativity. It is undeniable that his equations reveal a curious invariance of form for the Lorentz transformations, and that the wave-functions which are their solutions enable us to define magnitudes having a covariance in accord with the demands of Relativity principles: but serious difficulties still remain. First of all, in order to obtain results in agreement with relativistic covariance, Dirac had to introduce the possibility of the electron being in states of motion with negative energy, so that everything occurs as though the mass of the electron were negative: and further examination of this difficulty shows that the existence of such states of negative energy is intimately connected with the relativistic character of the Theory, so that it appears impossible to introduce the one without introducing the other. This difficulty, however, was eliminated at least in part by the discovery of the positive electron; but another and more important one remains. As has already been observed, the mathematical symbols defined by the Dirac equations can be interpreted physically within the framework of the generalized formulation of the new Mechanics in terms of probability; and this interpretation, at any rate in its present state, assigns a specific rôle to Time, thus impairing the relativistic symmetry of the four space-time variables. Accordingly its introduction into Dirac's Theory involves the introduction of this absence of symmetry. In other words, so long as Dirac's Theory is regarded as an analytical formulation devoid of physical significance it can be treated as being in accordance with Relativity; but as soon as we try to infer from it predictions experimentally verifiable, we are compelled to use wave-functions to define the possible values of the observable quantities and their respective probabilities; and at present it is impossible to do this in a way which does not assign a privileged rôle to Time. I cannot here, however, elaborate this difficulty, since it would compel us to enter into the whole complex formalism of the various quantum theories. But I may say that on closer consideration this difficulty seems to arise from profound causes, such as the fact that there exists a privileged sense for the time variable, and the persistence of physical units in time; and I have already indicated that these two fundamental facts involve a certain dissymmetry between Time and Space even in non-quantal Relativity.

The analysis of the difficulties which the search for a reconciliation between Relativity and quanta encounters has led various authors to interesting conclusions. Quite recently, for instance, Einstein placed at the foundation of the Theory of Relativity a closely reasoned investigation of the methods of length measurement and the synchronization of clocks in a Galilean reference system. When this question is re-examined, as has been done in detail by Schrödinger, and the quantum of action is duly taken into account, it is found that Einstein's analysis is rigorously valid only for bodies of sufficiently great mass, since the existence of the quantum of action involves the impossibility of an exact determination of a length and a time simultaneously without causing a change in the motion to which this length or time refers. It is impossible to evade this perturbation, and the result is that there exists a minimum of length, and of time, which can be measured for any given body, these minima being greater in proportion as the mass of the body is less. It would therefore be possible to speak of space and time intervals having any desired minuteness only in the case of infinitely heavy bodies, and it is only for this limiting case that Einstein's analysis can be considered absolutely exact from the quantum point of view. In other words, the Theory of Relativity in its present state is, for Quantum Physics, merely a limit valid in the macroscopic sphere, valid, that is, in proportion as the quantum of action can be regarded as negligible. Ultimately, this is a consequence of the impossibility of effecting a rigorous separation between geometry and dynamics, an impossibility due to the Quantum Theory which I have already stressed, and of which Planck's constant (h) is the mysterious manifestation.

I shall conclude briefly, for there is at present no final solution available for the problem of bringing about a reconciliation between quanta and Relativity. The Theory of Relativity is the culmination of the older macroscopic Physics, while the Quantum Theory, on the other hand, has its origins in the investigation of the universe of corpuscles and atoms. Their origins being so widely different, it is not at all surprising that their reconciliation should be no easy matter. At the present moment, therefore, when the tumultuous development of quantum theories is still a recent event, it is natural that this reconcilation should not yet have been satisfactorily effected.

V

PHILOSOPHICAL STUDIES
ON
QUANTUM PHYSICS

I

CONTINUITY AND INDIVIDUALITY IN MODERN PHYSICS

(1) In Physics, as in every other branch of knowledge, the problem of continuity and discontinuity has existed at all times: for in this science, as elsewhere, the human mind has always manifested two tendencies at once antagonistic and complementary. On the one hand, there is the tendency which tries to reduce the complexity of phenomena to the existence of simple elements indivisible, and capable of being counted; a tendency whose analysis of Reality seeks to reduce it to a dust-cloud of individuals. On the other hand, there is the tendency based on our intuitive notion of Time and Space, which observes the universal interaction of things and regards every attempt to disengage definite individual entities from the flux of natural phenomena as artificial. The conflict between the continuous view in Physics, and its opposite, has existed through many centuries with varying fortunes, each gaining an advantage over the other in turn, and neither winning a definite victory. For the philosopher there is nothing surprising in this, since the development of theory in every sphere of intellectual activity shows him that, if pushed to an extreme and opposed to each other, the concepts of both the continuous and the discontinuous are unable to give a correct rendering of Reality, which requires a subtle and almost indefinable fusion of the two terms of this antinomy.

What gives a special interest to the question of the continuous and the discontinuous in modern Physics is the fact that, during the last few years, it has arisen in a particularly clear cut and also novel form. More definitely then ever before, the need has been realized of effecting a synthesis of the two opposed points of view, while at the same time the very real difficulties raised by this problem have led physicists to discuss questions which pass beyond the proper technique of their science, and merge with the general problems of philosophy.

(2) To begin our present study, we must examine the forms in which the opposite concepts of the continuous and the discontinuous have crystallized in the course of centuries as far as Physics is concerned. I shall also discuss the difficulties they have encountered at all times.

Let us begin with the discontinuous. In Physics, from the ancient philosophers onwards, a tendency towards the discontinuous has consistently manifested itself in the shape of atomic and of corpuscular theories. Their ultimate aim has been to reduce Matter to a mere agglomeration of indivisible elementary particles—i.e. to analyse it into absolutely distinct individual entities definitely and completely localized in Space.

But the weakness of such a view, when carried to its conclusion, appeared immediately. Like Leibniz's monads, these isolated elementary particles, devoid of extension, cannot act on each other at a distance, since by hypothesis there is nothing whatever outside them: the Space separating them really is a void. Neither can they act on each other on coming in contact (by collision, as it is called in Mechanics), because, being of the nature of points, they could not touch without being merged in each other. In order to build up a Physics with these elementary corpuscles, therefore, the champions of the discontinuous must pour water into the pure wine of their doctrine by introducing alien elements; and they can do this in a variety of ways, which I shall proceed to examine.

For example, they can imagine their particles to be like small balls, which can impinge on each other, as did the originators of the kinetic theory of gases. But in that case the particle must have extension, and the question immediately arises of what is contained *within* the particle. To this various replies are possible. One may say that the particle considered in the first instance was not really as simple as had been assumed for working purposes, and that it must henceforward be treated as a complex system formed of smaller particles, which is what contemporary physicists did when they substituted a miniature solar system consisting of a central sun having a positive electric charge, and surrounded by electrons for planets (Bohr's atom), for the simple and literally atomic atom of the ancients. But obviously there is here a vicious infinite, since the new elementary particles of which the original particle, now seen to have been falsely so called, is supposed to be formed, will be faced by the same questions and the same difficulties.

Again, instead of resolving the particle into smaller entities, its interior might be imagined as being a continuous medium, which is what has been done in Lorentz's Theory, where the electron is taken to be a minute sphere of negative electricity. But this amounts to the assumption that ultimate Reality is continuous, not discontinuous; and, still further, an explanation will have to be given of why the continuous substance forming the corpuscle obstinately refuses to be subdivided; why the corpuscle, despite the changes of fortune to which it is subjected, succeeds in remaining equal to itself, and in preserving its individuality. In the electron Theory this latter difficulty is particularly serious, since it has never been explained how a sphere of negative electricity can persist, since all its constituent parts ought to repel each other.

There is yet another way in which the attempt can be made to build up a Physics of particles: it is the method implied in classical Mechanics. Matter is then assumed to be formed of particles strictly of the nature of points—the material points of the text books on Mechanics. In this way the difficulties relating to the internal structure of the particles are avoided. Next, in order to allow interaction between the particles—to draw them out of their isolation—they are assumed to be centres of forces, and to attract and repel each other at a distance. It would be superfluous to recall how fruitful this point of view has proved in Mechanics and Physics from the time of Newton until the present day. But the very admission of forces operating between the corpuscles involves, at any rate by implication, a concession to the continuous. Actually, in classical Mechanics the field of force surrounding a material

point is defined by a *continuous* function—the potential function—
and it is the existence of this field of force which determines the
motion of the other material points immersed in it. Ultimately,
this means that the presence of a corpuscle modifies the properties
of the entire surrounding Space; in other words, the corpuscle is
simply the centre of an *extended* phenomenon. Thus by giving up
the idea of the discontinuous in its pure form we reach a kind of
compromise in which Matter appears as consisting of individual
entities having spatial extension, but grouped around a centre
having the nature of a point.

(3) Let us turn next to the various theories of continuity in
Physics. When considering a liquid or a solid, one generally has
the impression that it is continuous. So that it is natural, when
erecting a theory of material substances, to assume that they can
be divided arbitrarily into elements of infinitely small volume,
each containing an infinitely small quantity of Matter, and each
being subject to the action of elements having a volume infinitely
near to their own. In this way a mechanics of continuous media
can be formed, known as the Theory of Elasticity for solids and
as Hydrodynamics for fluids. We know, however, that physicists
have gradually reached the conviction that the continuous character
of solids and fluids is illusory, and that in reality they consist of
atoms in motion, while it is only the obtuseness of our senses which
prevents us from perceiving this ultimately corpuscular structure
of Matter, and causes us to suppose it continuous instead. But a
heap of very fine dust also looks homogeneous to us, though the
particles of which it is composed are infinitely greater than so many
atoms.

The fact is that a genuinely continuous physical view is not to
be looked for in the theory of material substances: rather, it appears,
must such a view be sought in the description of Light given by
the Wave Theory. The ancient philosophers, and later Newton
and the majority of eighteenth-century scientists, had adopted an
atomic theory of Light; but in the seventeenth century, as we have
found already, the Dutch physicist Huyghens proposed a totally
different conception. In his view Light was an undulation pro-

pagated in a continuous medium, the ether, the latter being supposed to penetrate every material object and to fill all the regions of Space which appear void to us. The discovery of the phenomena of interference and diffraction made by Young and confirmed by Fresnel and, later, the very able development, again by Fresnel, of the mathematical theory of waves and its triumphant verification by experiment brought about, in the first half of last century, a complete abandonment of the discontinuous view of Light and the general adoption of the Wave Theory. In this way a new branch of Physics—optics—came to be formed, in which the notion of the discontinuous seemed to be entirely left on one side. A homogeneous and continuous medium, the ether, propagates certain disturbances, which mathematicians represent by continuous functions satisfying a certain partial differential equation—the equation of the propagation of light-waves. At this stage it might have seemed for a moment that the continuous shows itself superior to the discontinuous, since it meets its own requirements and also allows us to build up a coherent physical theory in which the notion of the discontinuous finds no place at all, whereas the ideas based on the discontinuous must appeal, more or less directly and more or less openly, to the continuous. But fundamentally this superiority of the continuous is an illusion. For we must remember, first of all, that the foundations of the mathematical analysis of continuity, on which the entire theory of wave propagation is inevitably built up, could be established on an absolutely exact basis only by a sort of *rapprochement* between the continuous and the discontinuous —by the famous "arithmetization," in fact, of the analysis. But apart from this very general point, we can realize that a completely continuous theory of Light cannot be really satisfactory, since it cannot be reconciled with the discontinuous structure of Matter. I shall stress this, a fundamental, point, because we shall here touch the roots of the curious critical period which has been affecting Physics for a number of years, a crisis to which the name of "quantum crisis" might be given.

When a disturbance is propagated in a continuous medium, it has a natural tendency to grow weaker as it spreads. Accordingly, on Fresnel's view, the Light given out by a source having the character of a geometrical point spreads uniformly over a sphere, whose radius increases proportionally with time in such a way that the density of the luminous energy diminishes indefinitely. The wave motion is thus disseminated, and the capacity for action grows steadily weaker. There is nothing here resembling the preservation of an individual entity, or the transference of a finite and localized power of action. Further, if the medium is genuinely continuous, by which I mean that if it is regarded under its most microscopic aspect, it is not resolved into an agglomeration of particles, then it is possible to show that the disturbances will have a natural tendency to be transformed into increasingly rapid vibrations, taking place on an increasingly smaller scale.

The purely continuous view of natural phenomena would thus lead us to expect the disappearance of all individual entities, and a tendency towards a homogeneous state in which energy would evolve towards increasingly subtler forms. Now Matter, in fact, reveals itself to us in the form of simple or compound chemical substances having immutable properties. We find too that electricity consists of elements—the electrons—which are invariably like themselves. It follows that the theorist cannot possibly represent the properties of Matter by the help of pure continuity alone. On the other hand, experiments seem to compel the acceptance of the undulatory aspect of Light, with the result that some thirty years ago the following compromise had been reached. Matter was supposed to consist of corpuscles which were immersed in the luminous (or the electromagnetic) ether. These corpuscles were further believed to be capable of acting on each other through the mediation of disturbances propagated through the ether, while when a material corpuscle was in rapidly accelerated motion, the continuous surrounding medium received an impulse which it transmitted in every direction in the form of a wave which grew weaker the farther it spread (Light, X-rays and other radiations).

Unfortunately this compromise, which assigned discontinuity to Matter and continuity to the luminous ether, was not destined to survive. To say nothing of Einstein's bold theories, which threw strong doubts on the very existence of ether, the simultaneous existence of a discontinuous Matter and a continuous ether was, as we shall proceed to show, logically impossible.

The position is, then, that if the propagation of Light in a vacuum is really strictly comparable to that of a disturbance within a continuous medium, then the vacuum will be able to absorb an unlimited quantity of energy, which will be transformed into vibrations of a continually growing rapidity and subtlety. Hence, if material bodies are placed within a closed region, an equilibrium of energy between these bodies and the surrounding vacuum will be impossible; and since continuous media have this unlimited capacity to absorb vibrations, the ether will absorb the entire energy of the Matter, which will therefore tend towards the state of immobility which physicists denote by absolute zero. No great perspicacity is required to see that this conclusion is absurd; and experiments show that it is in fact incorrect. In an enclosed region in a state of thermal isolation a temperature equilibrium invariably supervenes in the end; when the vacuum has absorbed a certain amount of energy provided by the Matter present it is as it were saturated; and the Matter conserves the remaining surplus. Further the vibrations in the vacuum do not continue to increase indefinitely in frequency: the fact is that the energy is distributed among the various frequencies of vibration in a perpectly stable manner defined by what is known as Planck's Law of black body radiation.[1]

It will now have become clear why physicists were unable for long to remain satisfied with the somewhat heterogeneous theoretical structure they had erected by the juxtaposition of a continuous Physics of radiation and a discontinuous Physics of Matter. It is here that the quantum crisis arose, whose chief characteristics I shall now attempt to set out.

(4) Since it is impossible to save the rigorously continuous view of luminous phenomena, one of the first forms in which quanta appeared was naturally that in which the idea that Light consists of individual entities is restored to respectability. The existence of light-corpuscles, of photons, was once again assumed.

[1] cf. p. 232, Appendix I.

Like the corpuscles of Matter, photons were supposed to be defined by their energy and their momentum, and also to preserve their individuality in moving through space. This hypothesis was strengthened by undoubted facts. There was first the photo-electric effect, whose proper character I shall shortly explain. If a light-source, as Fresnel supposed, emitted a spherical wave into the surrounding ether, the emitted energy would be scattered through space and the influence which the Light could exercise would become weaker with the distance. On the corpuscular Theory, however, the case is altered completely. For the corpuscle con-serves its energy, and whatever its distance from its source, it can always produce the same effect, in exactly the same way in which a shell full of explosive has the same destructive effect, whatever the distance from the gun whence it was fired. Now the photo-electric effect consists essentially of the fact that the effect of Light on atoms is the same at whatever distance from the source. This is an exactly ascertained fact, and it appears to suggest with unescapable force that luminous energy is concentrated in a cor-puscular form. Other phenomena, notably the diminution of the frequency of X-rays by scattering, discovered in 1922 by the American physicist H. A. Compton, seem to lead to the same conclusion.[1]

On the other hand, the correct interpretation of interference and diffraction seems to require the undulatory idea, with the result that since the discovery of the photo-electric effect the Theory of Light has been in a state of acute crisis. It looked as though the simultaneous introduction of the continuous and the discon-tinuous, in a completely unintelligible form, had become inevitable.

The quantum crisis has also had a profound influence on the Theory of Matter, since it was realized that the motion of material particles on a very small scale—e.g. that of the electrons within the atom as envisaged by Bohr—did not obey the laws of classical Mechanics. Among the infinitely numerous and continuous series of states of motion regarded as possible by classical Mechanics, only a restricted number, among which integers find a place, were

[1] cf. p. 233, Appendix II.

supposed to be stable. Here again discontinuity reappeared, but in a completely unexpected form, for what had become discontinuous was the possibilities of motion and not merely the structure of Matter. Within the atoms the fact seemed to be that the internal state remained stable for fairly lengthy periods, and was then suddenly transformed into another stable state. At the same time it seemed very difficult, and even impossible, to describe this "transition." The question might fairly be asked whether our customary ideas of Time and Space, that perfectly continuous framework within which we look at things, were genuinely valid as a description of the events taking place within the atom.

This rapid survey shows the importance of the quantum crisis within contemporary Physics. It is a severe crisis, and it has shaken the entire ancient structure of our scientific knowledge. In every field it has confronted with each other the two antithetic concepts of the continuous and the discontinuous, resuscitating the one where the other had hitherto triumphed. It is one of those stages where the human mind is forced to realize, with some bitterness, that the complexity of Reality refuses to allow itself to be cast in too simple a mould, and that a new and painful effort will be required for a fresh definition of what, after all, is perhaps indefinable.

(5) The time has now come to examine the fundamental idea which, elaborated first of all by the author, has been the foundation of the new Theory known today as Wave Mechanics.

This fundamental idea is that it is essential to introduce *simultaneously* the notion of the corpuscle and that of the wave in every branch of Physics alike, in the Theory of Matter as well as in that of Light. Every corpuscle must be regarded as accompanied by a certain wave, and every wave as being linked to the motion of one or more corpuscles. Sound reasons for accepting this point of view can indeed be found even in the older Mechanics, whence it is possible to deduce the relations which ought to subsist between the characteristic magnitudes of the corpuscle and the magnitudes which define its associated wave. We can thus see how it can be possible for the existence of light-corpuscles to be reconciled

with the way in which luminous energy is distributed in the phenomena of interference and diffraction. Thence we are led to the
assumption that phenomena analogous to the interference of
Light ought to be observable with material particles, with electrons
for example, and this wholly unexpected prediction has found
remarkable and quantitative verification in the discovery of the
beautiful phenomenon of the diffraction of electrons by crystals.[1]

Wave Mechanics, still further, furnishes us with an interpretation
of the fact that there are certain privileged stable states within
atomic systems. For since a wave must be associated with every
corpuscle, it follows that if the atom is regarded on Bohr's lines
the motion of the planet-electrons must be associated with wave
propagation. Mathematically, therefore, the atom will be similar
to a vibrating system; and we know that a vibrating system—e.g.
a vibrating cord, sounding pipe or wireless aerial—is generally
the scene only of certain privileged vibrations, namely those
corresponding to its *Eigen*-periods. Accordingly the atom too will
have its *Eigen*-periods, and will contain only waves having those
periods. This is the reason why, as Schrödinger has shown in
detail in his brilliant Papers, the series of possible stable atomic
states is discrete.

Without unduly stressing the development of Wave Mechanics,
I wish to make clear the difficulties which arose when an attempt
was made to define exactly the apparently inevitable introduction
of the notions of the corpuscle and of the wave simultaneously.

No doubt the simplest idea would be to treat every physical
entity as a periodic phenomenon, spatially extended, but centred
on a point. The motion of this extended entity would then be
equivalent to the propagation of a wave; not however of a homogeneous wave like those of Fresnel's optics, but of a wave having
a singular point, which would be the corpuscle in the narrow sense
of the word; in this way the corpuscle would be genuinely incor-
porated within the wave. Just as in the case of the material point of
classical Mechanics, surrounded by its field of force, the synthesis
of the continuous and the discontinuous would in this case be

---

[1] cf. p. 234, Appendix III.

brought about by the idea of individual entities having extension, but at the same time organized around a centre; and it would then be possible to hope to preserve in its purity the concept of Causality, together with the accepted schema of phenomena within the framework of Space and Time, or, if it is preferred, of relativistic Space-Time.

Unfortunately this method is impracticable and must be discarded, principally because the waves which the new Mechanics envisages are continuous and homogeneous like those of Fresnel; they have no singular point, and consequently there is no reason for saying that the corpuscle should be placed at one point rather than at another. As we go still more deeply into the problem, we realize that the concept of waves does not enable us to define the motion of the individual associated corpuscles, but that it only gives a statistical representation of their possible motion.

The question accordingly arose whether it were really possible any longer to regard the corpuscles as individual physical entities, completely defined and definitely localized in space; and today the reply given by theoretical physicists is in the negative. Following the ideas first put forward by Heisenberg and Bohr, they now hold entirely novel theories on this subject, of which I propose to give a rapid outline.

(6) In classical Mechanics, the material point is regarded as being quite definitely localized in space at every instant, i.e. as having at every instant unmistakably definite co-ordinates. Further, its state is completely described in terms of its energy and momentum; these magnitudes follow from its velocity, and they too are assumed to have a definitely fixed value at any given instant. If then we know that at the initial instant, t_0, its co-ordinates, its energy and momentum had a certain value, and if we also know the field of force to which it is exposed, then the laws of Mechanics enable us, at some later instant, t, to predict its position, and the value of its energy and momentum. Hence this old point of view implies a strict Determinism for mechanical phenomena. In practice, however, there invariably arises indeterminateness that is in a sense accidental, since it is due to the imperfection of our methods of

measurement. Actually, the co-ordinates and the initial velocity of any moving body are never known with absolute precision; all that we can say is that they fall within certain limits, generally very narrow, which give the exactness with which these magnitudes have been measured. From this slight indeterminateness of the initial data, however, there follows an indeterminateness in our predictions of the positions and final velocities of the moving body, this indeterminateness generally increasing with the lapse of time. But I repeat that according to classical notions this indeterminateness is contingent, so that it ought to be eliminated completely if only we succeeded in progressively perfecting our methods of measurement.

Now the new concepts introduced by contemporary physicists are the following. Beginning from the idea that every observation necessarily introduces some degree of disturbance into the phenomenon under investigation, they conclude, on the basis of an acute analysis, that even if we possessed infinitely perfect measuring instruments, it would be impossible to ascertain simultaneously and with absolute exactness both the position and the velocity of a corpuscle; quite apart from the contingent indeterminateness already referred to, there would always be an essentially irremovable indeterminateness. On the other hand, a corpuscle is detected only through its reactions with other corpuscles, e.g. with those forming our instrument of measurement. Every corpuscle, again, can be so detected only at intervals and at varying distances; whence it may be inferred that it is impossible, or at any rate superfluous, to assign a continuous trajectory and an instantaneous velocity to any corpuscle. The new Mechanics, therefore, is compelled to confine itself to connecting together the various successive and discontinuous manifestations of the corpuscle's existence, without attempting to state exactly what happens to it in the interval.

What then must be the methods of the new Mechanics? Let us assume that the corpuscle has been observed for the first time at the instant t_0. In that case the two types of indeterminatenesses already described will enable us to say only that at that initial instant the corpuscle was within a certain region of space and that its energy and momentum lay within certain limits. We shall thereupon construct, by methods which I shall not detail here, a wave which will represent this indeterminateness of our initial data. Next we shall calculate the propagation of this wave in space. Knowing the wave-form at an instant, t, later than t_0, we can predict that if at the instant, t, the corpuscle were detectable a second time, then the probability that it lies within a given region of space, and that its energy and momentum fall within certain given limits, is such and such.

We may thus consider the corpuscle "free," in a sense, to manifest itself at a variety of different places with an energy which may also vary; but, given the initial data, we can calculate *exactly* the probability that it will choose certain values and positions rather than others. The probabilities in question, still further, have the rather striking property of being capable of representation and calculation in terms of wave propagation "in space." As German writers say, there is a "wave-like transference of probability."

Certainly there is a marked element of paradox in these waves of probability, pure abstractions as they are which can nevertheless be propagated "in space" just as though they were elastic waves, and these corpuscles whose existence is intermittent while their exact localization and definition are impossibilities. But the fact that such concepts are today accepted by physicists clearly shows in how acute a form the problem of the continuous and the discontinuous has arisen in contemporary Physics, and how great are the difficulties raised by the fact that a synthesis between them must be brought about.

A serious objection may appear to stand in the way of the ideas just outlined. For if they are correct, how is it that the mechanical phenomena occurring on our own familiar scale appear to be governed by an absolute Determinism? To this objection, however, there is a good answer. If we take the macroscopic phenomena in whose case classical Mechanics can give exact figures, and calculate numerically the essential degree of indeterminateness introduced by the new ideas, we shall find that the latter is invariably very much smaller than the accidental indeterminateness due to the lack of accuracy in the measuring instruments. In these circumstances, the essential indeterminateness is completely masked by the errors introduced in the course of experiment, and everything happens therefore as though it did not exist at all. In other words, each corpuscle at each of its manifestations has always, so to speak, to make the choice between several possibilities; but the limits of this choice are supposed to be so narrow that in practice, as also in experiment, everything happens as though instead of a free choice there were a strict Determinism. Thus the *apparent* Determinism of large-scale phenomena by no means contradicts the ideas of the new Mechanics.

On the other hand, for phenomena on the atomic scale the essential indeterminateness becomes so great that a spatio-temporal description of states of motion becomes completely impossible. This is the explanation of the kind of uneasiness felt by physicists imbued with classical ideas on trying to form a picture of intra-atomic events.

(7) To complete this outline of the surprising concepts which have arisen in Physics, it remains to say something of the new statistical Mechanics associated with the name of an eminent young Italian scientist, Fermi. In the old kinetic theory of gases, a gas was imagined to consist of individual molecules, which were in competely orderless agitation whose average intensity determined the temperature of the gas. When two molecules are very close to each other they exert a mutual action on each other; they collide with one another incessantly, and these impacts are the means by which a certain distribution of energy between the molecules arises and is maintained (Maxwell's Law). These inter-molecular forces are active only at very minute distances, so that when two molecules are at an appreciable distance they do not act on each other in any way whatever, and their states of motion may be regarded as completely independent. Let us for example imagine a large receptacle with two molecules of gas at its opposite ends; then the fact that one of these has a certain energy can tell us nothing about the motion of the other. Such, at any rate, is the

classical assumption, which sounds like the very quintessence of common sense.

The development of Wave Mechanics, however, has led Fermi to a very different theory of gases. According to this new theory, no two of the individual entities forming the gas can be in the same state of motion and have the same energy[1]—a view which at first sight must certainly appear paradoxical. How, indeed, can it be possible for an atom at one end of a vessel containing gas to prevent another atom at the other end being in the same state of motion as itself? It seems inexplicable, except on the assumption that in a certain sense each individual entity composing the gas fills the whole of the receptacle.[2]

It appears that Fermi's theory is certainly valid for certain gases—those formed of electrons for example; in the interpretation of the electrical and thermal properties of metals, too, it has led to very striking results (Sommerfeld). Hence it must contain part of the truth, and we are once again led by it to the concept of physical entities indefinitely localized, an idea much less simple than that of the corpuscles and the atoms of the older Physics.

(8) Let us sum up in a few words.

Reality cannot be interpreted in terms of continuity alone: within continuity we must distinguish certain individual entities. But these individual entities do not conform to the idea which pure discontinuity would give us of them: they have extension, they are continually reacting on each other and, a still more surprising fact, it appears to be impossible to localize them and define them dynamically with perfect exactness at each instant. This conception of individual entities, rather vaguely outlined against the background of continuity, is something entirely novel for physicists, and seems to be a slightly shocking suggestion to some of them. Yet surely it harmonizes with the conception to which philosophical considerations might lead.

[1] This is a particular application of a principle formulated by Pauli and known as the Exclusion Principle; cf. p. 98.
[2] cf. p. 236, Appendix IV.

PLANCK'S LAW OF BLACK BODY RADIATION

Let us consider a vessel whose walls are maintained at a constant temperature and which contains, in however small a quantity, material substances at the same temperature and capable of emitting and absorbing radiation. In such a case Kirchoff has shown, by purely thermodynamic considerations, that the interchanges of energy between the material substances and the surrounding space, in which the radiation is situated, will lead to the establishment and maintenance of a state of equilibrium completely defined by the temperature. In this state of equilibrium the enclosed region should be filled with radiation in which all frequencies are present, and whose spectral distribution and total intensity are functions exclusively of the temperature. This radiation is known as Black Body Radiation characteristic of the given temperature.

From other considerations based on thermodynamics Stefan and Boltzmann succeeded in showing that the total energy of black body radiation, at absolute temperature T, is proportional to the fourth power of T.

Assuming the accepted ideas of some forty years ago (continuity of radiation and discontinuity of Matter), Lord Rayleigh found that the density $\rho(v)\,dv$ of the energy corresponding to a small interval of frequency dv in black body radiation at temperature T should be given by the formula:

$$\rho(v)\,dv = A v^2\, T\, dv$$

For a given temperature, experiments have shown that this Law holds good for small values of v, but not for large values.

If Rayleigh's Law were absolutely correct, it would lead us to attribute to black body radiation an infinite density of energy, as becomes apparent if we integrate $\rho(v)$ for all the possible values of the frequency. This corresponds to the fact already mentioned, that the entire energy of Matter ought, according to the classical theories, to be absorbed by the ether.

To obtain a more consistent Law than Rayleigh's, Planck assumed that Matter emits and absorbs radiation in finite quantities —by quanta. On these new assumptions, which in a sense imply the existence of a discontinuous structure of radiation, we reach the following new Law of spectral distribution:

$$\rho\,(\nu)\,d\nu = \frac{C\nu^3\,d\nu}{e^{\frac{h\nu}{kT}} - 1}$$

where k is the gas constant referred to a molecule.

For a given temperature Rayleigh's Law is found to apply for the low frequencies, while for the higher frequencies Planck's Law diverges from Rayleigh's and tends to the limiting form:

$$\rho\,(\nu)\,d\nu = C\nu^3\,e^{-\frac{h\nu}{kT}}\,d\nu$$

suggested by Wien before Planck's researches, and experimentally verified for high frequencies.

It is, further, easy to see by an integration of the expression $\rho\,(\nu)$ given by Planck that the total density of black body radiation is finite and is proportional to T^4, as required by thermodynamics.

APPENDIX II

THE COMPTON EFFECT

If we assume the existence of light-corpuscles, the encounter between an electron and a photon must be regarded as a collision in the general sense of the word. Let us assume that the electron is at rest; then if a photon strikes it it will be set in motion, while the photon will be diverted from its path to some extent. Owing to the Conservation of Energy the photon, having thus transferred some of its kinetic energy to the electron, will have lost some energy in the collision. According to the new corpuscular Theory of radiation, the frequency of the photon must be regarded as proportional to its energy, and accordingly the collision between a photon and an electron at rest ought to result in a lowering of the photon's frequency.

More specifically, if a beam of Röntgen rays falls on Matter, the X-photons collide with completely or nearly immobile electrons; the photons are then diverted from their original path, and at the same time undergo a lowering of frequency; this is linked with the deviation by a simple formula. This is described as X-ray scattering, and the scattered radiation has a frequency lower than that of the incident radiation, and varying with the angle of scattering. The actual existence of this fine phenomenon was first proved by H. A. Compton's experiments, and since then the proof has frequently been repeated; all the details of the phenomenon appear to agree with the predictions following from the simple theory of collision between electron and photon. For a full study of the subject technical works should be consulted.

APPENDIX III

THE DIFFRACTION OF ELECTRONS

When a wave passes through a medium in which centres are regularly distributed which are capable (a) of vibration under the influence of this primary wave and (b) of emitting secondary waves, there are certain directions in which the secondary waves will be in phase, and have maxima of intensity. But it is an essential condition of this phenomenon that the distances between the centres within the medium in question shall be of the same order of magnitude as the length of the primary wave. If this condition is fulfilled, part of the energy of the incident wave is scattered under the influence of the medium having this regular structure; but the energy thus diverted from its rectilinear propagation is concentrated in certain directions which are defined by the conditions of agreement in phase.

In natural crystals, still further, the molecules are arranged regularly at distances of the order of one Ångström unit (10^{-8} cm.). The wave-length of X-rays is of the same order of magnitude, and hence, if a beam of X-rays is directed on a crystal, we should expect to find that in certain directions there are maxima of intensity of scattering. To have predicted and proved the existence of this phenomenon (in 1912) was the great triumph of von Laue, and the later development of X-ray spectrography is based entirely on it.

Ultimately the Laue effect is just an interference phenomenon, and as far as X-rays were concerned its existence seemed natural enough, since their undulatory nature was taken for granted: they were assumed to be Light of very high frequency. The great merit of Davisson and Germer, and later of G. P. Thomson, was that they tried to repeat similar experiments by directing electrons against a crystal. If, in accordance with the new theories, the motion of an electron is to be regarded as closely linked with the propagation of a wave, it becomes natural to believe that part of the beam of electrons passing through the crystal will be scattered and that the electrons, thus diverted from their straight path, will be concentrated in certain directions defined by the conditions of agreement in phase in exactly the same way as for radiation. By different methods the physicists referred to succeeded in producing this phenomenon in 1927, and proved that the length of the wave associated with an electron does actually agree with the general formulae given four years earlier by the present author. Today the diffraction of electrons by crystals is an everyday laboratory experiment: it is used, not to demonstrate the principles of Wave Mechanics—these are today admitted on all hands—but to investigate the structure of bodies partially or wholly crystallized. It has also proved possible to obtain the diffraction by crystals of other particles than electrons—for example of protons and of various atoms. Here again technical works should be consulted for a deeper study of these questions.

SPATIO-TEMPORAL LOCALIZATION AND SPECIFICATION OF ENERGY

In Wave Mechanics it is no longer possible to state exactly and simultaneously the spatio-temporal localization and also the specification of energy (in terms of energy and momentum) of any given corpuscle. It is interesting to apply this idea to the Fermi statistics, already referred to. In the theory of gases the individual entities forming the gas are always assumed to be in an exactly determined state of motion; for example, in stating Maxwell's Law of distribution we speak of the number of molecules having a certain velocity or energy. But in the new Mechanics the spatio-temporal localization of these individual entities becomes impossible, and we can understand how it is that Fermi's Theory fails to provide a representation of molecular interaction within the framework of our habitual space. Light is thus thrown on the difficulty discussed at the end of the Chapter.

2

THE DIFFICULTIES OF DETERMINISM

Chaldean shepherds, stretched by the side of their flocks on quiet nights, are supposed to have been the first of mankind to follow the movement of the stars across the vault of heaven. They found that these movements are no matter of chance: they obey unalterable laws. Perhaps, contemplating the magnificent spectacle of the mighty celestial clock, their dim minds conceived the rudiments of a still more general idea, and came to suspect that Nature obeys Laws.

To say that there are Laws in Nature, then, means that phenomena are interconnected by an invariable order, and that, given the actual existence of certain conditions, the existence of a given phenomenon necessarily follows. As men passed from the stage of the Chaldean shepherds and learned better to observe the surrounding Universe, they began to discover in the physical world a growing number of Laws found to be invariably fulfilled, and their belief in the existence and immutability of physical Laws steadily grew. Thus in the minds of people who devoted themselves to the study of the sciences there became implanted more firmly the belief that the physical world is an immense machine, turning in a manner exactly determined in such a way that a complete knowledge of its state at a given moment would enable all its future states to be predicted. This theory of a rigorous and universal Determinism was laid down particularly by Laplace in his *Essay on the Calculus of Probability*, in which that great geometrician wrote the words justly famous for the exactness of the idea, and the elegance with which it is conveyed: "An intelligence, knowing at a given instant all the forces operating within Nature, and the respective position of all the entities of which it is composed, and further possessing the scope to analyse these data, could comprehend within one formula the motion of the greatest bodies within the Universe, and of the least atom; nothing would be indeterminate for it, and present and future alike would be present before its eyes."

For the mathematician, the Determinism of natural phenomena is expressed by the fact that these phenomena are governed by equations whose solutions are completely determined for all the time-values, provided that the values of certain magnitudes are known at a given initial instant.

In practice, the belief in Determinism has rendered a great service to scientists by preserving them from sloth. For whenever a scientist discovers a new class of phenomena of a confused and irregular appearance, he is tempted to surrender to indifference and discouragement, and to feel that these phenomena have no laws and that no profit can be derived from their study. But here his belief in Determinism intervenes to tell him that there must be laws not yet known to govern these newly discovered phenomena, and that, if once they were known, he could disentangle the confused skein of facts. So he returns to his labours, and often makes useful discoveries.

But the doctrine of Determinism has more than practical utility: it is certain that it contains part of the truth, since, if it were completely wrong, there would be no order nor regularity in physical phenomena, and no scientific knowledge of them could exist. But Physics does exist: it is an actual fact, and has proved its value by its progress and the number of its practical applications.

At the same time the notion of a universal and rigorous Determinism undeniably raises a fair number of objections. There is, for example, the vital question whether it leaves its rightful place to that teleological activity which is manifested in living Nature. Again, does it give its due to the spirit and its manifestations in the world of actuality? I shall not venture here on a discussion of such deep questions. My aim is a more modest one: what I wish to do is to explain the crisis from which the idea of Determinism in Physics has been suffering for a number of years. Physics—by which I mean the science of dead Matter—had up to the point in question seemed to be the very stronghold of Determinism, and even the opponents of that principle had seemed to be ready to leave this field to it. Yet the most recent theories, adopted by physicists almost against their will to explain facts experimentally observed, lead, not so much to a complete surrender of Determinism in Physics (I have already said that the simple fact of the existence of Physics does not allow of this), as to the view that it is not complete nor universal, and that in fact it has limits. What I wish to attempt to explain here is why and how this unexpected change has been brought about in scientific thought.

*   *   *

The framework of our perceptions is three-dimensional Space, and we have a tendency to assume that the whole of physical Nature ought to be capable of exact representation within this framework. But our perceptions are not unchangeable: they undergo modification in the course of time. Accordingly the natural tendency of people who tried to construct physical theories has been to regard the Universe as being formed of elements having at each instant a certain arrangement in space—a distribution changing through time, as is clear from the fact that transformation takes place in the physical Universe. Accordingly, the state of the physical Universe at a given instant is supposed to be completely defined by the distribution of these elements, i.e. by a certain configuration, or "figure" as they would have said in the seventeenth century; and the evolution of the material Universe was supposed to correspond to the progressive variations of this configuration. It was for this reason that Descartes, in an attempt to chart the future course of modern Science, wrote that we should attempt to explain physical facts "by figures and motion." Such a concept is completely in accord with the principle of Determinism, if it is the fact that a knowledge of the position and velocity of the elements of the physical world at a given instant is sufficient to determine completely their later motion.

The most perfect type of explanation in accordance with the Cartesian ideal is supplied by the corpuscular theories. In these theories the assumption is that Matter consists of corpuscles, or material points, i.e. of minute, simple and indivisible elements whose extension is so slight that they can be treated as identical with geometrical points. The spatial distribution of these corpuscles, and their motion in time, are supposed to account for the properties of Matter. The first question which now arises is as to the character of these corpuscles, and how many different kinds of them we must suppose to exist to give a successful account of Reality. During the last century, chemical discovery taught us that all chemical substances are derived from the combinations between certain "simple substances," reaching the respectable number of 92. Chemists were thus led to the assumption that each simple substance consists of so many atoms, each of them identical with all the others. Physicists adopted this view of the atom and constructed a number of theories (the most familiar being the Kinetic Theory of gases) in which the atoms were cast for the part of elementary corpuscles. But it was impossible to halt at this point; the scientific spirit, in its craving for simplicity, could not remain satisfied at a stage where it had to operate with 92 different species of elementary corpuscles. The experimental discovery that electricity had a corpuscular structure came at this point to simplify matters. Experiments showed negative electricity to consist of corpuscles all of the same (extremely slight) mass and electric charge—the electrons. Since then it has been proved that positive electricity also has a similar corpuscular structure, the elementary corpuscle of positive electricity being the proton. At this point physicists realized that the atoms of simple substances must not be taken to be elementary corpuscles, but were complicated structures formed of protons and electrons,[1] there being 92 different types of these structures—and 92 different kinds of atoms. A great step had thus been taken towards a simple corpuscular Theory of Matter, since the two classes of corpuscles by themselves would suffice to account for the properties of Matter and to reduce the entire material Universe to a vast collection of protons and electrons. If, further, it were to prove possible to find exact

[1] But cf. p. 76.

laws governing the motion of these corpuscles, then the Cartesian ideal of a description of the physical world in terms of "figures and motion" would have been fulfilled, and simultaneously the demands of the doctrine of universal Determinism would have been met. It looked as though physicists were on the point of reaching an ideal pursued for many years.

*   *   *

Let us recall the intellectual standpoint of a physical theorist of some thirty years ago. For him, Matter was reduced to a collection of protons and electrons, and the essential, indeed one might say the sole, problem was to know what Laws of Motion must be applied to these corpuscles. Accordingly it was natural to take the Laws of Newton's classical Mechanics as those which govern the motion of corpuscles; and in fact this Mechanics found excellent support from the study of the motion of the celestial bodies and of those material objects which surround us on the Earth's surface. Hence it seemed permissible to apply it by extrapolation to the ultimate elements of which Matter appeared to be constituted. Now one of the essential characteristics of Newtonian Mechanics, as applied to corpuscles, is that it is determinist. On the classical view, in other words, the corpuscle is a simple material point of negligible dimensions; at every instant it has a position in space which is exactly determined, and during the flux of time it describes a curve known as its trajectory. Given the position and the velocity of a corpuscle at a certain instant, therefore, the equations of classical Mechanics enable us to predict exactly the entire future motion of the corpuscle. Certainly, if classical Mechanics were really applicable to material corpuscles, and if at any given instant we could know exactly the positions and velocities of the vast number of corpuscles which between them form the material Universe, then the entire future course of this material Universe would be rigorously determined, and the ideal defined by Laplace in the words quoted above would be attained, at least in principle. But here I must insist on one point: the exact determination of the motion of a corpuscle is based essentially on the assumption, tacitly made by classical Mechanics, that it is possible to know exactly—i.e. to measure precisely—the position of a corpuscle and, simultaneously, its state of motion as defined by its velocity.

Some years ago, the deterministic doctrine was triumphant in the various corpuscular theories, but since then it too has found that the Tarpeian rock is near the Capitol. For while the application of the laws and conceptions of classical Mechanics to the ultimate elements of Matter led at first to encouraging results, it was found in the last analysis to be incapable of accounting for actual conditions as revealed by experiment. The reason for this reverse was the discovery of a new class of phenomena—quantum phenomena —which it is impossible to interpret in terms solely of the classical concepts. I cannot here describe quantum phenomena, and the difficulties of their interpretation, in detail: all I wish to do is to stress two aspects of these difficulties. On the one hand, as I have explained in earlier Chapters, the study of quantum phenomena has led physicists, after lengthy gropings, to assume that the properties of Matter cannot be explained by the hypothesis that it consists of corpuscles only, but that waves must be associated with these corpuscles, the physical significance of these waves, which we shall examine further below, being of a surprising kind. On the other hand, the formulation of the laws based on the experimental data about quanta has invariably brought in the new constant, of which classical Physics knew nothing, called Planck's constant. In equations this constant is usually represented by the letter h; it cannot be interpreted within the framework of the classical theories based on Newtonian Mechanics; in fact, its meaning, for the last thirty years and largely even now, has been the enigma of modern Physics; and it has remained the hidden syllable within Nature's cross-word puzzle.

One extremely important point is here to be noted. The value of the constant h is extremely small relatively to the magnitudes which are usually found in phenomena on the human scale. It is for this reason that its existence drew attention to itself only when we had learned how to study the structure of Matter, which means the phenomena of the atomic and the sub-atomic scale.

* * *

The time has now come to indicate exactly the way in which contemporary theories have been led to associate waves with material particles. The first step in this undertaking must be to recall in the simplest possible terms what we mean by a wave. We can form an idea of a simple wave by imagining a series of undulations following each other at regular intervals, the distance between any two consecutive crests being known as the wavelength, and the height of the crest as its amplitude. The length and amplitude of a wave are the two magnitudes which define a simple wave. (Often such a simple wave is called a monochromatic wave, a term borrowed from optics.) But we can imagine more complex waves, formed by the superposition of monochromatic waves, and to describe a complex wave of this kind we must have the lengths and amplitudes of all the constituent simple waves, or as it is called by analogy with optics, we must have the "spectral decomposition" of the complex wave.

I said above that, after lengthy research into quantum phenomena, physicists reached the conviction that the idea of a wave must, in the Theory of Matter, be associated with that of the corpuscle. The first discovery was that, given the motion of a corpuscle with a definite velocity, the propagation of a monochromatic wave ought to be made to correspond to this motion, the wave-length being connected with the velocity of the corpuscle by a relation containing the constant h. This was the rudimentary idea whence developed the new Mechanics known as Wave Mechanics. In Wave Mechanics, the study of the motion of a corpuscle is *replaced* by the study of the propagation of the associated wave. This propagation of the associated wave obeys exact laws; but it does not follow from this that an exactly determined motion can be assigned to the corpuscle, as I shall now endeavour to explain.

On the new view, as I have remarked previously,[1] it is the wave

[1] P. 188.

associated with a corpuscle that represents, or symbolizes, all that we know about the corpuscle. Generally the associated wave is a complex wave, characterized by a certain spectral decomposition and having a resulting amplitude which, at any given instant, is spatially distributed in a certain manner. Now the new Mechanics cannot assign to a corpuscle spatial positions that are continuously and exactly defined: all that it does is to tell us that the corpuscle must necessarily lie in the region occupied by the wave, and that its chance of being at a given point within this region is greater in proportion as the wave-amplitude is greater at that point. In the same way, the new method does not enable us continuously to assign to the corpuscle motion that is completely determined: for to each monochromatic component in the spectral decomposition of the associated wave there corresponds a possible value of the corpuscle's momentum, and all that we know is that the momentum is defined only within certain limits.

Thus in the new Mechanics there is always some uncertainty as to the position of the corpuscle, and also its state of motion. A study of the mathematical properties of the waves, however, soon shows that these two uncertainties are not independent: the smaller the one, the greater is the other. To appreciate this, let us first examine the limiting case—that of a simple, monochromatic associated wave. We saw that this wave corresponds to a corpuscle whose momentum is known exactly. But it can be shown that a monochromatic wave of this kind has an infinite spatial extension and possesses the same amplitude everywhere, which means, in terms of the new Mechanics, that the associated corpuscle has a completely indeterminate position—that it may equally well be at any point in that space. Hence a complete knowledge of its motion involves complete uncertainty as to its position. It may, however, happen that the associated wave, instead of being infinite, only fills a limited region R, outside of which its amplitude is zero. In this case the uncertainty regarding its position is smaller, since it is certain that the corpuscle must be somewhere within the region R. But mathematical analysis shows that a wave, thus limited to a given region in space, must be complex: it is formed by the superposition of monochromatic waves, each of which corresponds to a possible momentum of the corpuscle. Here, then, the uncertainty as to position is no longer complete; but to make up for this we have uncertainty as to motion. Finally, we may consider the limiting case at the other end of the scale—that of a wave occupying an infinitely small region R. In this case the position of the corpuscle is accurately defined, since it must be within R; but then, a wave of this kind, having infinitely small dimensions, can result only from the superposition of monochromatic waves of all possible wave-lengths; whence it follows that the corpuscle may have any one among all possible momentum values. Thus as soon as uncertainty about position ceases, there is complete uncertainty about momentum.

Werner Heisenberg, who was the first to realize these consequences of the new Mechanics, has expressed them mathematically by means of equations known today as Uncertainty Relations. These bring out the fact that it is the existence of the constant, h, which prevents us from knowing simultaneously and exactly both the position and the motion of a corpuscle, while if h were zero, such simultaneous knowledge would be possible.

It might be objected that simultaneous knowledge of both the position and momentum of a corpuscle could be obtained by a simultaneous measurement of these two magnitudes. To this objection Heisenberg has triumphantly replied that there is no process of measurement or observation which could possibly enable us to ascertain, simultaneously and exactly, the position and the momentum of a corpuscle. For any apparatus permitting measurement of position brings about—in a way we do not understand—an interference with momentum, and does so the more intensely as measurement of position becomes more accurate. And conversely, any apparatus enabling us to measure momentum affects the position in a way we do not understand, and does so the more powerfully as measurement of momentum becomes more precise. A closer examination of the question thus brings us back, through this critique of the possibilities of measurement, to the Uncertainty

Relations—the Principle of Uncertainty—already deduced from the properties of associated waves.

*   *   *

In the new Mechanics therefore, as we have seen already, we can never assume that both the position and the initial momentum of a corpuscle are known simultaneously. The consequence is that there can be no absolute Determinism. Actually, the predictions of the new Theory are very different from the exact predictions of classical Mechanics. If we wish to start from an initial state, in order to predict later states, we must observe that the initial state cannot be known exactly, since there are inevitable uncertainties regarding the position and momentum of corpuscles. The initial state will be represented, complete with the uncertainties which it involves, by a certain initial form of the associated wave. The later changes of the wave can be predicted exactly by the equations of Wave Mechanics: but this does not mean an absolute Determinism for the corpuscle, since knowledge of the wave at every instant only enables us to assign certain probabilities to the various hypotheses tenable as to the position and momentum of the corpuscle. In a word, then, while the older Mechanics claimed to apply exact and inexorable laws to every phenomenon, the new Physics only gives us laws of probability, and though these can be expressed in exact formulae, they still remain laws of probability. Thus in every physical phenomenon there remains a margin of uncertainty; and it is possible to ascertain that this margin of uncertainty is measured, in a way, by the constant, h. Figuratively, it has been said that there was a crack in the wall of physical Determinism, whose size was measured by Planck's constant, h; and thus the latter receives a somewhat unexpected interpretation: it is supposed to be the limiting barrier of Determinism.

One objection, however, naturally arises. The mechanical phenomena on our own familiar scale, and also on the astronomical scale, seem to follow a rigorous Determinism, and, as I said at the beginning, it was this very fact which suggested the principle of a universal Determinism. How then can this Determinism of macroscopic phenomena be reconciled with the ideas which I have just advanced? The reply is simple if we remember the minuteness of the constant, h, relatively to the magnitudes operating in mechanical phenomena on the human, and *a fortiori* on the astronomical, scale. I have already observed that the margin of uncertainty in these phenomena, measured by h, is so very minute as to be negligible, and is further entirely masked by the experimental errors which inevitably affect our observations and measurements. Hence there is an *apparent* Determinism in macroscopic phenomena, which in no way conflicts with a certain indeterminateness in phenomena on the microscopic scale.

There are, however, many physicists who look back with longing to the absolute Determinism which they consider essential for the progress of Science, although many others have been less reluctant to give it up. The former hope that in some way or other it will be possible to resuscitate it; the others have their doubts.

In any case, in the present state of our knowledge the Cartesian ideal of representing the physical world by means of "figures and motion" seems to have suffered bankruptcy. After all, Heisenberg's Principle of Uncertainty, with its prohibition of an exact and simultaneous knowledge of position and velocity, is surely the very expression of the fact that it is impossible to know, at the same moment and exactly, both "figures and motion."

3

THE NEW IDEAS INTRODUCED BY QUANTUM MECHANICS

In modern times, from the eighteenth century to the present day, theoretical Physics seems essentially to have rested on two foundation stones: the mechanical view of the material World which had already been formulated by Descartes, and the theory of the continuity of physical phenomena, which allowed their investigation to be undertaken with the powerful help of infinitesimal analysis—that theory which was so clearly expressed in Leibniz's famous formula: *natura non facit saltus.*

The brilliant successes of these two concepts in numerous scientific fields (Astronomy, Mechanics, and Physics proper) had led the majority of physicists, or at least those who did not take refuge in a drastic formalism, to adopt more or less implicitly a certain simple view of the physical Universe. They imagined it to be like a gigantic machine which, as Descartes had put it, could be described by means of Form and Motion, i.e. as a machine consisting of parts in spatial juxtaposition which could move in Time, while their motion and the changes they underwent were essentially continuous and resulted from their mutual actions and reactions, so that they could be exactly expressed in the precise terms of mathematical analysis. As the interactions between the different parts of the material Universe diminish with distance, it is often possible to carve out independent or practically independent systems; and in such a case the Laws both of Mechanics and of Physics said that an adequate knowledge of the initial state of such an isolated system could predict exactly, given the help of analysis, the entire future history of the system. That such a prediction of the history of a fragment of the material World, if it is sufficiently isolated, is actually possible for the phenomena of the human and the astronomical scale is proved beyond dispute by the success of the great classical theories of pure Mechanics, of celestial Mechanics and of the mathematical Physics of last century. Thus there was a time when it might fairly be thought that the same concepts of an all-pervading Mechanism would lead to successes of the same type if applied to phenomena on the atomic scale. For this reason the attempt was made, once the existence of elementary corpuscles of Matter, of electrons and protons, had been experimentally discovered, to represent phenomena on the atomic scale by the motion of these elementary particles, and to treat the atoms as though they were a kind of planetary system, reproducing on a small scale the planetary systems of Astronomy. It was found necessary, however, to introduce quantum conditions into this planetary model of the atom, and this fact proved that in reality a factor intervened on this scale which was entirely negligible on the larger scales. This factor (already referred to) is the quantum of action or, if it is preferred, Planck's constant, h; and we have seen that Bohr, whose name had previously been given to the planetary atomic model, first of all suspected that this simile was actually misleading, and that the intervention of the quantum of action on the atomic scale set a limit to the applicability of the ideas of classical Physics. Later, too, it was pre-eminently Bohr who insisted most on this point, in proportion as the new quantum theories developed. I shall therefore follow his exposition and try to show how the theory of an all-pervading Mechanism on the one hand, and that of continuity on the other, were breached by the development of the new Mechanics.

*   *   *

To understand one of the principal reasons why mechanism was checked in this way, we must examine the position of the physicist who observes and measures in order to calculate and predict, on the assumption that such a physicist is working within a universal Mechanism and a material Universe assumed to resemble a vast machine. In all the classical theories, as I remarked in the previous Chapter, there is a hypothesis made more or less implicitly—viz. that it is possible to observe and even to investigate quantitatively the state of a given system without interfering with this state. In other words, it is assumed that a scientist can observe a system, and even undertake measurements within it, without any appreciable exchange of energy between the system and himself. Now it is quite clear that this is not strictly true even in the case of simple observation, since we cannot have any acquaintance with the external world except through our senses, and since further any sensation assumes that there must be some action of the external world on one of our sense organs, and consequently an exchange of energy between the external world and our body. Hence *a fortiori* in the case of an experiment or a measurement where the physicist actively interferes with a definite object in view, it is not strictly permissible to neglect the interaction between the system under examination and the apparatus used by the physicist who is examining it. At the same time this theoretical objection can be neglected in practice whenever we investigate a phenomenon on the macroscopic scale, i.e. on the human or the astronomical scale. It is quite certain that when the astronomer turns his telescope on a star in order to observe it he affects the star's orbit to an entirely negligible degree; and even with the phenomena occurring on the Earth around us a skilful experimenter can always succeed in avoiding any appreciable interference with the phenomenon he wants to investigate. But the transference of the physical mechanism from the macroscopic sphere, where it certainly is valid, to smaller and smaller regions where measurement has to be applied to smaller and smaller quantities, assumes in effect that it is possible to reduce indefinitely the interaction, implied in any measurement, between the external world on the one hand and the apparatus or senses of the experimenter on the other. Otherwise, there must come a moment when the interference due to measurement becomes comparable in magnitude to the measured quantities, in which case it will be impossible to determine what is the actual value of the magnitudes, after measurement, which it is desired to measure. It is at this point (as I have already re-marked) that the existence of the quantum of action comes into play, or Planck's constant, h, the appearance of which was one of the most essential factors in the development of contemporary Science. From the fact that this constant exists, and has a finite value, it follows that in a system on the atomic scale it is impossible to measure simultaneously and exactly all the magnitudes which must be known exactly and simultaneously if we wish to have an exact mechanistic description of the system. More precisely, when there are two canonically conjugate magnitudes, in the sense of analytical Mechanics, like the co-ordinate x of a corpuscle, and the component of its momentum along the axis of x, then these two magnitudes cannot be measured exactly and simultaneously. The more carefully an experimenter conducts his measurements so as to obtain a more exact value of one of these magnitudes, the more, by this very operation, and without any possibility of avoiding it, the other magnitude is affected by the very act of measurement, and affected in a way which is unknown. These facts, as I have previously remarked, have been demonstrated by the subtle and profound analyses made by Bohr and Heisenberg. Thus the existence of the quantum of action introduces into every act of measurement a finite disturbance which cannot be checked, and which affects every pair of canonically conjugate magnitudes. Such is the physical content of Heisenberg's Principle of Uncertainty. It should be noted still further that, strictly speaking, this impossibility of measuring exactly all the magnitudes necessary for the mechanistic definition of a system also applies to macro-scopic systems, but that the minimum uncertainty of the con-jugate magnitudes is of the order of Planck's constant, and can here be rendered entirely negligible relatively to the magnitudes involved in such a system, so that the mechanistic conception is very approximately valid. To state this more exactly, we shall denote as macroscopic a phenomenon of such a kind that Planck's constant can be treated as infinitely small relatively to the mag-nitudes involved in the phenomenon, and we shall denote as microscopic a phenomenon occurring on so minute a scale that it becomes impossible to disregard the finite value of the constant, h. Hence in the microscopic sphere, thus defined, the reaction between the observer and the external reality, which necessarily accompanies every observation and every measurement, profoundly affects the extremely delicate phenomena which it is desired to investigate. Hence Heisenberg's Uncertainty; hence too the impossibility of predicting exactly, in the case of microscopic phenomena, the future development of a system starting from a known initial state: since all the magnitudes which it is essential to know simultaneously to make such a precise prediction cannot be simultaneously known.

A problem here arises which has frequently been stressed by Bohr and Heisenberg. Classical Physics assumes that there exists an objective Reality, which can be described absolutely independently of the "subjects" who observe it; indeed this is the very point, as Bohr has well observed, on which classical Physics was able to base its claim to be an exact science. But now it is found that in modern microscopic Physics we can no longer make a clean distinction between the phenomenon observed or measured, and the method of observation and measurement. The microcosm, therefore, is not an objective Reality which man can conceive or describe independently of the instrument which acquaints him with it. Either we make no attempt to follow the changes of the microcosm, in which case they remain unknown and are no object of scientific research, or else we do try to follow them, in which case we influence them in a way which we cannot check, and which in fact depends on the very elements which we wish to discover exactly. Bohr has inferred from all this that Quantum Physics reduces or blurs the dividing region between the subjective and the objective; but there is possibly some misuse of language here. For in reality the means of observation—the instruments of measurement and even our sense-organs—clearly belong to the objective side; and the fact that their reactions on the parts of the external world which we desire to study cannot be disregarded in microscopic Physics neither abolishes, nor even diminishes, the traditional distinction between subject and object. If then Bohr's idea is to be expressed in rather more valid terms it should be done, I consider, in the following way. Classical Physics makes an artificial cut between one part of the objective world, which it calls external Reality, and which is completely independent of the "subjects" who observe it, and another part of the external world, viz. the instruments of measurement and the sense-organs which are supposed to serve the "subjects" in the process of becoming acquainted with, and studying quantitatively, the external Reality but without ever influencing and modifying this. Quantum Physics, on the other hand, shows that such a cut is artificial, and demonstrates that a description of physical Reality which is wholly independent of the means by which we observe it is, strictly, an impossibility. Thus the new microscopic Physics can at best claim to effect a connection (and we shall see that even then its predictions are of a merely statistical nature) between one set of facts experimentally discovered and another, later, set of facts, in which each discovery introduces unknown modifications. Hence the degree of causal determinateness, the existence of which in the objective world can be demonstrated by scientific research, is proportionally diminished.

Following out his arguments, then, Bohr has remarked that the kind of disturbance introduced by observation into the phenomenon to be observed in microscopic Physics has a certain similarity to the difficulty met with in Psychology, when it is desired to make an objective study of psychological phenomena by introspection. For the great difficulty which the psychologist has to overcome when practising introspection, a difficulty which prevents the results of his investigation from ranking as an exact science, consists in the impossibility of concentrating his attention on a mental process without by that very act modifying, or even completely arresting, the process itself. To take an instance belonging rather to the sphere of Physiological Psychology, if we try to observe introspectively the psychological phases accompanying the transition from being awake to sleep, the result is generally disappointing: nothing at all is observed, because the observer does not succeed in falling asleep; and the attention which it was desired to concentrate on the gradual process of falling asleep has prevented this phenomenon occurring.

* * *

What has been said should, I think, suffice to make clear the way in which the mechanistic view of Physics came to clash, in the microscopic sphere, with the fact that it is impossible to outline the processes occurring in the physical Universe without somehow introducing the disturbances due to observation and measurement. At present my aim is to show how the other fundamental concept on which classical Science rested—I mean that of the continuity of natural phenomena, by virtue of which it was possible to apply Infinitesimal Analysis to them—has equally been shaken by the progress of Quantum Physics.

I have stated already that from the fact that the constant h has a finite value, there followed the impossibility of knowing simultaneously and exactly the values of two canonically conjugate variables. Now what are these pairs of variables described as being canonically conjugate, when regarded from the physical point of view? If we examine these pairs of variables, we shall find that there is always one which serves for the description in terms of Space and Time of the system under examination, while the other serves for the specification of its "dynamical state." Heisenberg's Principle of Uncertainty asserts that owing to the finite value of the constant, h, we cannot know simultaneously and exactly the pairs of conjugate variables, which means (in other words) that the complete description of a mechanical system in terms of Time and Space is incompatible, in current Science, with the exact specification of the dynamical state.

The impossibility of giving an exact description of the spatio-temporal localization, and of the dynamical state, simultaneously, may perhaps be connected with one of the difficulties which troubled ancient philosophers. Let us take an arrow in flight, said Zeno. At any given moment it is motionless in a certain position. How then can it follow a certain trajectory? How—that is to say—

can motion be constructed out of a series of immobilities? To the eyes of modern Science, imbued as it was, before the discovery of quanta, with the idea of continuity, the argument from Zeno's arrow used to look rather childish. Consideration of the positions infinitely close to each other occupied by a body in motion does in fact enable us to define the velocity by the same argument as that which serves to establish the existence of the derivative of a continuous function, and velocity is ultimately just the derivative of position with respect to time. In classical Mechanics, then, the co-ordinates of a moving body at a given instant t, or more generally of a mechanical system, define its spatio-temporal localization, while the derivatives of the co-ordinates with respect to time define its state of motion. Consideration of positions infinitely close to each other, and of the instantaneous uniform motion which permits the transition from one of these to the other, thus seems to open the way to a complete refutation of Zeno's objection. But this refutation itself rests on the assumption that physical phenomena are continuous, and is impaired by the introduction of an element of discontinuity into the physical Universe like that implied in the existence of the quantum of action. Without wishing, then, to make Zeno a precursor of Heisenberg, and without forgetting the part played in this problem at the present day by the finite value of Planck's constant, we may still say that the impossibility, which recent theories reveal, of assigning simultaneously to a moving body an exact spatio-temporal localization and also a completely defined dynamic state, appears to have some kinship with a philosophical difficulty which has long been familiar. To use Bohr's terminology, the exact localization in Space and Time is one "idealization," and the concept of a completely defined state of motion is another, so that the two "complementary idealizations," while almost quite compatible on the macroscopic scale, are not strictly so on the microscopic scale.

The consequences of the surrender of continuity, which were implied in the introduction of the quantum of action, were clearly revealed when Bohr's Theory of the atom was formulated in its original version. In this Theory the atom is taken to be a miniature solar system. At present the search is proceeding for the exact quantitative data of this solar system. But the introduction of the quantizing of the atom, which was found necessary to harmonize theory with facts, compelled Bohr to suppose that the solar system in question could only assume a certain number of "stationary" states, the energies of which form a discrete series. The passing from one stationary state to another is accompanied by the emission of radiation, and it has to be regarded as a sudden transition eluding description. Here, then, for the first time in modern Science, the idea became apparent that there is a fundamental incompatibility between the concept of a stationary dynamical state, characterized by a definite energy-value, and a description in terms of Space and Time. This incompatibility was found to be even deeper when the development of the new Mechanics showed that the positions and the velocities of the electrons (the planets in Bohr's miniature solar system) have no real existence, and that actually it is not only the transitions from one state to another, but also the stationary states themselves, which elude every attempt at spatio-temporal description. The stationary state—i.e. the stable and exactly defined energy-state—is an idealization which, on the atomic scale, is no longer compatible with that other idealization implied in the idea of a description within the framework of Space and Time.

Simultaneously with this, something like a kaleidoscopic character of the Universe considered in its "fine structure" was revealed. Successive observations of the same atom will always show it to us in one of its stationary states, without any possibility that we shall be able to seize it in the intermediate stage between two stationary states. Thus atomic Physics can apparently only reveal discrete values of energy, with the apparent consequence that it must renounce continuity and confine itself to the discovery of the laws—necessarily statistical—which govern the transitions from one value to the next. I say that these laws are necessarily statistical; for causal determination in Physics seems to be firmly connected with the idea that the transformations take place continuously within the framework of Time and Space, and appears to vanish with this idea. The new Physics then will try, given a system in an initial stationary state, to calculate the probability that the system will subsequently be found in such and such another stationary state.

The new conception—the result of the investigation of atomic stationary states with quantized energies—has been extended by recent mechanical methods to apply not only to energy but also to all the other measurable physical magnitudes. The exact measurement of any one of these magnitudes always leads to a definite value; but the values found for one and the same magnitude, by successive measurements, are interconnected solely by laws of probability. To understand this question, we must now consider the way in which the "dualism" of corpuscles and waves has entered into modern Physics. The more successful, then, investigators were in improving their methods of research in the atomic and microscopic sphere, the more clearly it appeared that the results of their experiments are most naturally expressed on the assumption that the structure of physical reality is discontinuous, discrete units being distinguished within it, some of them complex like the molecule and atom, and others simple (at least provisionally) like the proton, electron and photon. The result of an exact measurement is invariably formulated by the assignment of certain values to certain of the magnitudes which define these discrete physical units. As far as Matter is concerned, it has long since been established that the results of these delicate experiments always reveal either corpuscles, or independent systems of corpuscles, like atoms; and still more recently, the analysis of the reactions between Matter and radiation has shown that the case is the same for Light, and has drawn attention to light-corpuscles, or photons.

Accordingly, all the results of experimental microscopic Physics are now stated in corpuscular form: a fact which cannot be evaded. But the elementary corpuscles, or the independent systems of corpuscles which experimental research enables us to isolate in the domain of physical Reality, can never be completely described if we follow the classical method of exactly localizing them in space at a given instant and simultaneously assigning to them a precisely defined dynamic state. This fact is connected, as we saw, with another—viz. that it is impossible to observe Reality without introducing perturbations, as expressed by Heisenberg's Principle of Uncertainty, which follows from the existence of the quantum of action. But since we never know all the necessary data, from the point of view of universal mechanism, for predicting the future changes of a corpuscle or an atom, the question arises as to what kind of prediction, or if it is preferred, of Science, remains possible. It is at this point that the concept of waves intervenes; for as I have repeatedly observed, the development of Wave Mechanics has caused a wave to be associated with every corpuscle, or system of corpuscles; and the character of the wave has gradually been shown to be somewhat symbolic. Each measurement gives us information about a corpuscle, or corpuscular system, which is inevitably incomplete because of Heisenberg's Uncertainty Principle; but after any such measurement we are able to associate with this corpuscle (or system) a certain wave which in a way represents our knowledge, including the uncertainty from which it suffers. The future changes of this wave, as time passes, can then be followed exactly by means of the propagation equations of the new Mechanics; and the wave-form at a given later stage enables us to predict the possible results of measurement of any given mechanical magnitude made at that stage, as well as the respective probabilities of the different possible results. In this way the development of the probabilities after the first measurement follows a strict Determinism, which is symbolized by the propagation of the wave; but it does not follow from this that there is in general a strict Determinism of the corpuscular phenomena which are directly discovered by microscopic investigation. Further, since each new measurement introduces a new and unknown factor, the entire calculation of the wave changes must be undertaken afresh after each measurement; nor is this surprising, since the wave represents the probabilities, and obviously each measurement disturbs the probabilities merely in supplying us with the new information.

It is, in short, the corpuscular aspect of microscopic phenomena that is always experimentally detected; and if it were possible to measure simultaneously all the factors involved in the classical notion of the corpuscle, we should no doubt reach a mechanistic and deterministic description of the microcosm, analogous to that aimed at in Bohr's original theory. Today, however, we know that a simultaneous knowledge of the spatio-temporal, and of the dynamic, factors is made impossible by the existence of the quantum of action; and the result is that the succession of observable conditions, when applied to the idea of the corpuscle, appears discontinuous and non-causal. The notion of waves undeniably permits continuity and Determinism to be restored as between two consecutive measurements; but this applies solely to the future course of probabilities. The new Mechanics, by taking its stand on ideas of this order, has incidentally found it a simple matter to explain how it is that we find apparent continuity and Determinism in the macroscopic sphere, where the constant, h, is practically negligible, and also how it is possible, in that region, to reconcile spatio-temporal idealization with the idealization of dynamic states. In other words, it has proved possible to show that when we pass from the microcosm to the macrocosm, Quantum Physics tends asymptotically towards classical Physics.

I shall make one more observation. When the history of a single physical unit is investigated, whether corpuscle or atom, the associated wave symbolizes the probabilities of localization and of the dynamic state of that unit; but when we are dealing with a very great number of identical units, then the wave represents the statistical distribution of this totality of units. Let us take, for example, the experiments in classical optics. Regarded from our present point of view, these experiments are found to operate with a great number of photons, though the actual existence of no single one of them is revealed. All that they enable us to do, therefore, is to investigate the statistical distribution of the photons, and this distribution is represented by the associated wave, which is none other than the classical light-wave. This is the ultimate reason why classical optics has been able to dispense with the concept of the corpuscle, and to concentrate solely on the wave. The same conditions are also encountered in recent experiments on the diffraction of electrons. The results of these experiments can be described solely in terms of the wave associated with the electron, there being no scope for the concept of the corpuscle. Thus too we come to understand why, despite its symbolical character, the waves can appear to be physical Reality itself, in certain experiments involving a great number of corpuscles.

*   *   *

Let us sum up the ideas which we have pursued hitherto. On the microscopic scale, physical Reality is found to consist of units which undergo successive transformations, together with sudden transitions. These transformations, however, cannot be described by means of infinitesimal analysis within the framework of continuity and Determinism. On the other hand, the statistical aspect of these kaleidoscopic transformations can be described in classical scientific terms by employing the artifice of the associated waves. When on the other hand we pass to the macroscopic sphere, where Planck's Constant no longer has any appreciable effect, the discontinuous character of these individual phenomena vanishes, submerged, so to speak, under the flood of statistics. At the same time the complementary and more or less irreconcilable descriptions in terms of corpuscle and wave, of spatio-temporal localization and of dynamic state, are combined and merge in the strict and harmonious mould of classical Physics.

It may be worth while to observe that the existence and finite value of Planck's Constant introduce an essential difference between the microcosm and the macrocosm. The idea that the physical world is like itself on every scale without exception, and that the infinitely small is a mere small-scale reproduction of the infinitely great, is found as a *Leitmotif* in the writings of thinkers and the theories of scientists. The genius of Pascal expressed this idea in deeply moving terms; and twenty years ago we found it at the root of the simile of the planetary atom. Today, however, in the light of recent theories, this conception must be regarded as incorrect. Ideas which suffice for the description of the macrocosm are inadequate for that of the microcosm; and when the physicist, descending the scale of magnitudes, comes to the atomic world—the world of corpuscles—he finds there an entirely new and irreducible element, the quantum of action, which brings with it the important consequences which I have attempted to analyse. The finite value of Planck's Constant permits a definition of the microscopic scale, and of its opposition to the macroscopic; with the result that in the world of Physics the infinitely small ceases to be the mere small-scale representation of the infinitely great.

*   *   *

Must we go still farther, and must we believe, as Bohr appears to suggest, that the new ideas of contemporary Physics will allow us to understand why it is that the classical methods of objective Science do not seem to adapt themselves with a good grace to the phenomena of the vital and of the mental order, with the result that, according to Bohr, microscopic Physics would be the intermediary between macroscopic physical Reality, where mechanism and determinism are valid, and the other and subtler spheres, where the same concepts are, if not wrong, at any rate inapplicable? As a physicist, I shall not attempt to reply to this question. The sole conclusion which I shall express is that the discovery of quanta, the consequences of which are only now beginning to become apparent in all their scope, seems to require of scientific thought one of the greatest changes in orientation which it has ever had to make in its secular effort to adapt as closely as possible our idea of the physical world to the requirements of our reason.

4

THE SIMULTANEOUS REPRESENTATION OF POSSIBILITIES IN THE NEW PHYSICS

In the two preceding Chapters I have explained the deep change which the development of the new physical theories, due to the study of quantum phenomena, has imposed on classical views. It is a subject of fundamental importance, and also one of considerable difficulty; so that it seems worth while to pursue it once again from a slightly different point of view than that taken up previously. If we wish to appreciate the complex architecture of a large building, we grasp the harmony of the whole by examining every aspect of it which lies open to us.

*
*
*

To approach our subject by a gradual path, we shall begin by assuming the intellectual standpoint of the physicists of the end of the nineteenth century, and consider the Wave Theory of Light as it presented itself to them.

Accordingly, we assume provisionally that Light is a disturbance propagated in a medium which penetrates all bodies. It is a highly hypothetical medium, and to disguise our ignorance of its exact nature we begin by calling it the ether. The disturbance of the ether which constitutes Light will be represented at any given point by a directed magnitude (the light-vector), which is equal to the displacement of the element of the ether situated at this point with reference to its position of equilibrium. This displacement is further assumed to occur in a direction perpendicular to the direction in which the disturbance is propagated; for since the time of Fresnel we know that light-vibrations are transverse. To simplify all this still further, let us assume that all the displacements are parallel, in which case we say that the light is polarized. The displacement of the element of the ether at any given point will then be stated in terms of its elongation alone, and this will be a certain function, generally complicated, of the time: $f(t)$. A complete description of the luminous disturbance requires a knowledge of all the $f(t)$ functions relative to all points of space. In more mathematical terms, the function of the space and time variables, which represents the entire luminous disturbance, is decomposed into an infinite number of functions of the time, each relative to a point in space. The disturbance at any given point will vary in its intensity with the magnitude of the displacement at that point of the element of the ether; and in the classical Theory of Light, it is possible to show that the luminous energy present at any point is proportional to the square of the function $f(t)$ relative to that point. Thus we see that by considering the decomposition of the function, which represents the entire luminous wave, into functions $f(t)$ associated with the various points—that is, the "spatial decomposition" of the wave-function—we immediately obtain the spatial distribution of the luminous energy.

But there is one particularly simple class of luminous disturbances —plane monochromatic waves; and when a plane monochromatic wave passes through the ether, there pass through it regular waves, the crests and troughs of which succeed each other at regular intervals. This equal interval between the successive crests is the wave-length, λ, of the monochromatic wave. Further, at every point in the ether there is a periodic vibration with the period T; and then knowledge of the wave-length λ, of the period T, and of the direction of propagation, defines the plane monochromatic wave. Naturally, however, the simplicity and regularity of the plane monochromatic wave give it a quite exceptional character, and generally a disturbance cannot be reduced to such a plane monochromatic wave. At the same time, consideration of this simple type of wave is of fundamental importance in the Theory of Light on account of Fourier's famous Theorem: applied to luminous disturbances, this theorem tells us that all of them can be analysed into a superposition of plane monochromatic waves. To state this in precise mathematical terms, the function of the space and time variables, which represents the entire luminous disturbance, can always be decomposed into the sum of a finite or infinite series of sinusoidal functions, each of which represents a plane monochromatic wave; and in this way we obtain the "spectral decomposition" of the wave-function, which is wholly different from the "spatial decomposition" to which I have already referred. For we no longer assign to each point a function $f(t)$ representing the luminous vibration at that point, however complex it may be; we now consider simultaneously a totality of plane monochromatic waves, whose superposition is equivalent to the entire actual wave. At first sight this new decomposition of the wave seems much more artificial than the first, since what seems to be the actual entity, so long as we believe in the existence of the ether or, more generally, of "something vibrating," is the vibration (which generally is complex) of this "something" at each point; while the operation by which the totality of the local vibrations is decomposed into a superposition of monochromatic waves seems to be purely intellectual. This impression, however, is weakened by a little reflection, since we know that optical apparatus, like prisms or lattices, can actually decompose a complex luminous wave—white sunlight, for example—into monochromatic waves: by means of such apparatus, therefore, the various monochromatic waves are actually separated from each other, each with that amplitude which the application of Fourier's Theorem leads us to assign to it in the spectral decomposition of the incident wave. Thus spectral decomposition is seen to have a greater actuality than could at first be believed, since appropriate apparatus does in fact enable its monochromatic components to be extracted from the complex wave. Hence these components exist, in some kind of potential state, within the original complex vibration.

But what exactly is this potential existence? It is a question which attracted some of the greatest physicists of the classical period. Scientists of the stature of Henri Poincaré, Lord Rayleigh and de Gouy paused to discuss the real nature of white Light, and to give replies—divergent replies—to the following question: "Do colours exist within white Light before it is decomposed by the prism, or are they manufactured by the prism after the white Light has fallen on the latter?" We shall see below that contemporary Physics envisages this question from an altogether novel angle.

So far we have been assuming that the vibrations of the ether at all points were parallel. This is a hypothesis which simplifies matters; but if we abandon it we must regard the light-vector as capable of having, at every point of the disturbance, any orientation whatever within the plane perpendicular to the direction of propagation. Let us select two axes Ox and Oy, perpendicular to the direction of propagation and also perpendicular to each other; in that case we can decompose all the light-vectors into two components, one parallel to Ox, which at every point will be a function $f(t)$ of the time, and the other parallel to Oy, which at every point will be a function $g(t)$ of the time. The sum of the squares of the functions $f(t)$ will represent the total intensity of the component polarized parallel to Ox of the luminous disturbance, while the sum of the squares of the functions $g(t)$ will give the total intensity of the component polarized parallel to Oy. A Nicol prism suitably orientated could stop one of these components while allowing the other to pass. But there is an infinity of ways of choosing two rectangular axes Ox and Oy in a plane perpendicular to the direction of propagation, and consequently there is an infinite number of ways in which an unpolarized beam can be decomposed into two components polarized at right angles, which a Nicol prism suitably orientated will be able to separate. Here again it is clear that if we believe in the existence of the ether, or at least of something vibrating, it is the light-vector with its own specific orientation that is the physical reality at any given point. The decomposition of the light-vector into two rectangular components, which is possible in an infinite number of ways,[1] appears to be a purely intellectual operation even more than does spectral decomposition.

[1] Unpolarized Light can also be decomposed into components circularly polarized in opposite senses.

Yet at the same time, since a Nicol prism suitably orientated can extract from an unpolarized beam the component which is parallel to a given axis, it is surely possible to maintain that all the decompositions of the unpolarized beam into two components polarized at right angles do, in a potential sense, exist before the Nicol comes into action. It is a notion which brings with it a certain feeling of uneasiness so long as the framework of the classical conceptions is adopted, though it raises no difficulties whatever from the mathematical point of view. I imagine that many young physicists experience this uneasiness when they begin the study of physical optics. We shall see that here again the point of view of the new theories is very different from that of the old, and possibly the former is calculated, if not to dissipate the uneasiness to which I have referred, at any rate to explain its origin by assigning to it a deep-rooted cause.

*   *   *

Hitherto we have been making use of somewhat old-fashioned terms in speaking of the ether and its vibrations; the time has now come to prove ourselves more modern and to use less obsolete words; I shall therefore deal with photons.

The development of the Theory of Relativity has caused the idea of the ether—both the elastic ether of Fresnel and the electromagnetic ether of Maxwell—to be abandoned. At the same time the light-wave must be introduced for an interpretation of all physical optics; but though it has subsisted in Science, it has ceased to be the vibration of any entity. It is now merely defined abstractly by a vectorial function of the time and space co-ordinates which can always be represented, as has already been explained, by a system of variable vectors assigned to all points of space, though at present it is impossible to give to these vectors any clear and definite physical interpretation.

The concept of the photon, still further, entered the Theory of Light, where it completed the concept of waves by opposing itself to it. A detailed investigation of the elementary phenomena of the exchanges of energy between Matter and radiation, as in the photo-electric and the Compton effect, has shown that everything happens in these phenomena as though the luminous energy, and that of the other types of radiation, were concentrated in corpuscular form, not spread continuously in the wave as the classical Wave Theory required. Whenever we are dealing with a monochromatic wave with wave-length λ and period T, both exactly ascertained, the corpuscles contained in the wave (the photons), are found to have an energy equal to the product of the inverse of the period T and Planck's constant, h, and a momentum p equal to the product of the same constant, h, and the inverse of the wave-length λ. These are the Einstein relations:

$$E = \frac{h}{T} \qquad\qquad p = \frac{h}{\lambda}.$$

The problem is to reconcile this corpuscular structure of Light with the fact, proved a thousand times with the utmost exactness, that in the phenomena of interference and diffraction luminous energy is distributed as the square of the supposed luminous vibration at each point. It appears that the phenomenon becomes intelligible in one way only, which is by assuming that the luminous vibration of the classical theory, far from really representing the vibration of any actuality, is an abstract magnitude whose square gives the probability that a photon will be detected at the point in question. On this view, the function of the time and space co-ordinates representing the entirety of the wave gives us, by its "spatial" decomposition, the respective probabilities that the photon shall be present at different points of space. As the wave-function is generally different from zero in any extended region, there are different possibilities for the localization of the photon.

There is, however, another way of decomposing the wave-function—that is by "spectral" decomposition into a superposition of plane monochromatic waves. How are we to interpret this, while taking due account of the existence of photons? To reply to this question let us consider a beam of white light falling on a prism. Before the light reaches the prism we cannot assign any definite wave-length, i.e. any colour, to the photons, since this light is a mixture of different wave-lengths. It follows that we cannot make use of Einstein's equations to assign to the photon any definite energy and momentum. On the other hand, once the Light has been decomposed by the prism, we shall be able to assign a wave-length and a momentum to the photons which have now passed into the different monochromatic beams which, from this stage on, are separate. Ultimately, therefore, a prism (or a lattice) is an apparatus which enables us to assign to photons a wave-length and a momentum—or, if it is preferred, a frequency and energy. If for a moment we may use the language of metaphor, without at the same time forgetting that such a language has its dangers and must not be taken literally, we might say that the photon contained within the complex incident light—at which stage of its journey it has neither a well-defined wave-length nor momentum—is asked a question by the prism: "What is your wave-length?"; and the photon has to reply by choosing one of the monochromatic beams which leave the prism. Now the position is that the spectral decomposition of the incident complex wave represents the various replies which the photon can give, together with their respective probabilities, in the same way as that in which the spatial decomposition of the wave, into functions assigned to each point in space, represents the different possible localizations of the corpuscle and their respective probabilities.

We shall encounter the same considerations in dealing with the decomposition of an unpolarized light-wave into two components polarized at right angles. In the incident Light we cannot assign any polarization to the photon; but when it passes through a Nicol prism it is compelled, in a manner, to choose between one direction of polarization, and the direction perpendicular to this. Here again it is easy to see that the decomposition of the incident wave into two components polarized at right angles—which is possible in an infinity of different ways—represents the possible replies of the photon to the analysis made by the Nicol prism and their respective probabilities.

To sum up, we obtain the general idea that the light-wave, in all its possible decompositions, represents the total possibilities relatively to the photon associated with it.

*　*　*

Thus far we have been speaking of light-waves and photons; but all that has been said can be transposed from the sphere of Light to that of Matter. For the theoretical concepts which served as foundation to the lofty structure of Wave Mechanics, and the brilliant experiments on the diffraction of electrons by crystals, have shown that the corpuscles of which Matter consists must always be assumed to be associated with waves. The question then is: what are the relations which we must assume to subsist between the material corpuscle and its wave? The reply is simple: the relations are *exactly the same* as those which we were led to assume to subsist between the photon and the light-wave. Thus the wave associated with a corpuscle of Matter—an electron for example—is represented by a certain function of the space and time co-ordinates, and hence the "spatial" decomposition of this wave function into $f(t)$ functions, associated with the various points of space, will help us to predict the possible localizations of the electron in space; for here, as with photons, we assume that the probability of detecting the electron at a given point is proportional to the square of the function $f(t)$ at that point; and as I have already observed, the function of the wave is generally different from zero in an extended region, so that there will be more than one possibility for the electron's localization.

Wave Mechanics shows that the momentum, p, of a material corpuscle is connected to the length, λ, of the associated wave by the same equation, $p = \dfrac{h}{\lambda}$, which connects the photon's momentum to the length of the light-wave. Hence a well-defined wave-length can be assigned to a material corpuscle—to an electron for example —only on condition that the wave associated with the corpuscle shall be capable of spectral analysis into a superposition of mono-chromatic waves. Any apparatus designed to measure the momentum (or the energy) of the corpuscle is in reality an apparatus which separates the monochromatic components of the wave, just as the prism separates the components of the light-wave; and, like the latter, it compels the corpuscle to choose between its components. To resume the metaphor already employed, we may say that the measuring apparatus asks the corpuscle: "What is your momentum?" and the "spectral" composition of the incident wave before it reaches the measuring apparatus represents the various possible replies, as well as their respective probabilities.

In the early days of Wave Mechanics it was thought that the wave associated with the material corpuscle had a scalar character, and presented nothing resembling polarization. Later theoretical development has shown, however, that it was necessary, more particularly in order to explain the magnetic properties of the electron revealed by the investigation of the fine structure of spectra and other phenomena, to introduce a kind of polarization of the waves associated with the electron; and it was especially Dirac's Theory of the magnetic electron that made an exact interpretation of this polarization possible. It is a somewhat difficult problem, and I cannot pursue it in detail here:[1] but polarization of the wave associated with the electron is not identical in all respects with polarization of the light-wave. Yet here again, as in the case of Light, it is possible to say that the various ways of decomposing the wave associated with an electron into polarized waves, represent the different possible results of measurement that enables us to assign to the electron a definite polarization.

To sum up, we may conclude with regard to Matter as we did with regard to Light, that in all its decompositions the wave represents schematically the total possibilities relatively to the corpuscle associated with it.

*　　*　　*

I have now dealt with three types of wave decomposition— those which correspond to its position, to its momentum and to its

[1] cf. pp. 146, 158, 209.

polarization respectively. But it has also been possible to generalize all this, and to show that a certain decomposition of the associated wave corresponds to every measurable mechanical magnitude relating to the corpuscle, and that this decomposition gives the possible values of this magnitude and their respective probabilities. In this way a theory has been obtained at once comprehensive, exact and mathematically satisfactory.

One further important point should here be stressed. When we proceed to measure, with the appropriate apparatus, a magnitude, relating to a corpuscle, for example its position or momentum, then the new Mechanics tells us that if we wish to predict the result we must effect a certain decomposition of the associated wave. But it may happen that in effecting this decomposition we find that only one of the components differs from zero, in which case we say that the wave is simple with respect to this decomposition. For example, the wave-function may happen to be zero everywhere except at a point A, in which case it will have only one spatial component, and we shall be certain that the corpuscle will be able to manifest its presence only at the point A. It will thus be possible to assign a definite position to the corpuscle in advance. It may also be the case that the wave is a plane monochromatic wave; it will then have only one spectral component, and we shall be certain that measurement of the momentum will give us a certain value. We shall thus be able to assign in advance this definite momentum to the corpuscle. These are what theorists call "pure cases"—pure as regards position in the first instance, and as regards momentum in the second. But here an important question arises: can we have a pure case simultaneously for two different magnitudes relating to the corpuscle? The reply is that we can have such simultaneity for certain pairs of magnitudes, but not necessarily for all pairs; and mathematical theory enables us to ascertain the pairs of magnitudes for which this "pure case" simultaneity cannot arise. Now position and momentum are such a pair. In other words, a wave cannot be simple simultaneously with reference to its spatial decomposition and also to its spectral decomposition; if it is simple with reference to the one it must be complex with reference to the other. According to modern ideas, therefore, it follows that we cannot assign to the corpuscle definite position and definite momentum simultaneously. But it also follows that no measurement can enable us to measure simultaneously and exactly the position and the momentum of a corpuscle, since otherwise, after the measurement, the wave associated with the corpuscle, which represents our knowledge about the corpuscle, would necessarily be simple simultaneously both in its spectral and in its spatial decomposition: and we know that this is quite impossible. Thus we are enabled to understand the origin of the fundamental Uncertainty introduced into Physics by Heisenberg—a subject already discussed in my preceding Chapters; and by a still closer formulation of the argument we obtain the inequalities, familiar today, in which this Uncertainty expresses itself.

And now, to conclude, we can reply to a question such as the following: "Do colours exist in white Light before it passes through the prism which is destined to decompose it?" We shall reply that they do exist, but only in the way in which a possibility exists before the event which will tell us whether it has in fact been realized. It is a reply of some subtlety, and would certainly have surprised the physicists of yesterday: but it is also characteristic of the degree of keenness and abstraction which has been reached by physical theory today.

VI

PHILOSOPHICAL STUDIES
ON
VARIOUS SUBJECTS

I

PHYSICAL REALITY AND IDEALIZATION

THE physical world surrounding us is exceedingly complicated, and a constant process of abstraction and schematization has been required of scientific research to enable it gradually to carve out from Reality groups of phenomena capable of being united under a single theoretical figure. It has thus been found possible to isolate series of facts in the surrounding Reality, and to make them correspond to series of relations or of ideas logically connected with each other. In this way physical theory was constructed, and it is certain that its success has demonstrated the possibility of arranging, at least roughly, a large number of categories of phenomena within the framework of certain logical systems constructed by human reason. In a sense, this general agreement between external things and human reason is no small miracle, though it is to be observed that if it did not exist human life would be impossible, since in the absence of any relation between our understanding and facts we should be incapable of foreseeing the consequences of our actions. But if the progress of our science enables us to ascertain the details of phenomena with great exactness, thanks to a more refined experimental technique, does it follow that an exact correspondence must be maintained indefinitely between all the details we shall ever observe, on the one hand, and a perfectly defined logical framework on the other? Is it certain that the static concepts of our reason, with their clear-cut and bare outlines, can be applied so as to correspond exactly with a moving and infinitely complex Reality? Such questions have always been asked; but they seem to become more urgent now that the recent developments of Physics have overthrown a great number of our old ways of thought. I desire here briefly to consider these questions.

*   *   *

On more than one occasion Niels Bohr, whose penetrating thought has contributed so much to the advance of theoretical Physics during the last twenty years, has insisted that the existence of the quantum of action compels us today to employ "complementary" descriptions to account for the phenomena on the atomic scale. By this term we are to understand descriptions which are certainly complementary but at the same time, taken strictly, incompatible, the exact meaning of which postulate I shall explain in a moment. According to Bohr, still further, each of these complementary descriptions is an "idealization" permitting us to represent certain aspects of the phenomena under consideration, but not all the aspects.

The best known instance of such complementary descriptions is supplied by the two descriptions of Matter and Light by means of waves on the one hand and of corpuscles on the other. The employment of each idea, as we have abundantly seen, has proved essential for the interpretation of some phenomenon or other; but the two ideas still remain, despite every effort, incapable of being reduced to terms one of the other, and the only connection that can be established between them is of a statistical nature.

On further consideration, it is seen that the idea of the corpuscle is closely connected with that of spatial localization. The extreme and most exact, and as it were the purest, form of the concept of a corpuscle is that of a geometrical point to which a certain mass is assigned. In this form we have a limiting concept—a kind of extreme idealization. The position is that modern Quantum Physics tells us that we can localize a corpuscle in space, by appropriate experiments, with a degree of exactness which theoretically is unlimited; but it also tells us that the case of an *exactly* localized corpuscle is a limiting case, the probability of whose existence is always nil. In dealing with macroscopic phenomena, where things are observed on the large scale, the concept of a corpuscle as defined in a rather vague manner serves perfectly adequately for an interpretation of the facts. But if, by means of a more delicate process of observation, we descend to phenomena on the atomic and the microscopic scale, and if (further) we wish at the same time to give complete precision to the idea of the corpuscle taken as a mere point, we shall find that this idea, while not actually wrong, is none the less imperfectly suited to give a minutely detailed description of the observed phenomena.

Now the same set of facts can be discovered if we look at matters from the undulatory point of view. In classical dynamics there are certain magnitudes playing an essential part, viz. energy and momentum. The importance of the part they play results from the fact that there exist conservation theorems for these magnitudes which enable us to define, in the changes of a dynamic entity, certain characteristics which remain unchangeable. Accordingly a dynamic state is thus defined essentially by the values—which are constant—of the energy and the momentum. Now in modern Physics the idea of the dynamic state is associated with that of the plane monochromatic wave: this is the foundation of Wave Mechanics. Given a dynamic state defined by a certain value of energy and another of momentum, there corresponds to this state a plane monochromatic wave whose frequency and velocity of propagation are exactly determined; the result being that we find the conservation theorems as given in classical Mechanics under a new form. At the same time, if we examine the way in which the facts are presented to us in Wave Mechanics, we shall find that the concept of a well-defined dynamic state, associated with that of a strictly monochromatic wave, is an extreme idealization—a limiting case which is never fully realized in Nature. In exact terms, the wave associated with a material unit is never strictly monochromatic—it is always the resultant of the superposition of a number of monochromatic waves occupying a finite spectral interval. The notion of the dynamic state, satisfactory enough in macroscopic Physics, where matters are treated on the large scale, is found, as soon as things are regarded very closely, to be an idealization which never strictly fits Reality.

Thus we find that the corpuscle perfectly precisely localized at a certain point in space, and the dynamic state perfectly precisely defined and represented by a strictly monochromatic wave, are both abstractions which can, indeed, in certain cases correspond to the details of the observed facts with a considerable degree of exactitude, but which never render them with absolute and literal exactness. The curious fact is that, in each individual case, the one idealization departs from Reality in proportion as the other approaches exact agreement; qualitatively, this is the meaning of Heisenberg's famous Uncertainty Principle. We can now see the way in which the complementary descriptions by waves and corpuscles complete, by excluding, each other. They complete each other, in the first place, because we must appeal to each in turn according to the phenomenon which we have to describe, but, in the second place, they exclude each other because the better one of them is adapted to Reality, the worse is the other—and conversely.

*   *   *

It is easy to find other analogous circumstances in modern Quantum Physics. Let us take as an example the concept of a physical unit, and in particular that of a material unit like the electron. It cannot be disputed that in a great number of phenomena we can distinguish physical units, and can, for example, follow the actual path of an electron in a Wilson Cloud Chamber. Thus the concept of the individual physical entity can roughly be applied to Reality. But if we aim at greater exactness, and desire a rigorous definition of the individual physical entity, we find that we must take a unit completely severed from the rest of the world. For from the moment when interaction between a plurality of units commences, the individuality of each of them suffers a kind of diminution. This fact can be observed even in classical Physics: it is the concept of potential energy, by means of which the older theories symbolize action and reaction between physical units, and calculate the result. Now the potential energy of a mutual action belongs to the whole of the system, and cannot in any logical way be distributed between its constituent entities. Let us take a simple case of this—that of two electrified particles reacting on each other. Classical Physics assigns to each of these particles a kinetic energy, and the total energy of the system of the two particles is the sum of these kinetic energies plus the potential energy, which represents the interaction between the particles according to Coulomb's Law. Now this potential energy can be attributed to neither of the two particles alone, but belongs to the system. Thus in the classical theories the potential energy denotes the fact—in a manner which while obscure is also profound— that material units suffer a mutilation of their individuality when interaction takes place between them.

The same position recurs in a still more emphatic form in Quantum Physics. This has shown that there is a kind of complementarity between the concept of the individual unit and the concept of a system. In Quantum Physics, therefore, the system is a kind of organism, within whose unity the elementary constituent units are almost reabsorbed. When forming part of a system, then, a physical unit loses a large measure of its individuality, the latter tending to merge in the greater individuality of the system. This principle is particularly clear in the case of particles of the same kind, and is manifested by certain completely unexpected consequences which the classical concepts could never have led physicists to suspect, but which are in complete agreement with a large number of facts experimentally ascertained (the new statistics, the Exclusion Principle, etc.).

To make a real individual of a physical unit belonging to a system, then, it is necessary to take this unit from out of the system —to break the link which binds it to the total organism. If this is understood, we can also understand the way in which the concepts of the individual unit and of the system are complementary: the particle cannot be observed so long as it forms part of the system, and the system is impaired once the particle has been identified. The concept of the physical unit thus becomes completely clear and properly defined only if it is regarded as a unit completely independent of the rest of the world; but, since such an independence obviously cannot be realized, the concept of a physical unit taken in absolute strictness is found in its turn to be an idealization; in other words, to be a case which is never rigorously adapted to Reality. It may be mentioned that precisely the same applies to the concept of a system. By its strict definition, the system is an entirely enclosed organism completely unrelated to the external world, and therefore this concept is strictly valid only for the entire Universe.

At the same time, these notions of the physical unit and of the system are adapted to Reality and are useful for describing it, but only on condition that conditions are not analysed too carefully. But if we insist on perfectly exact definitions and, at the same moment, on a completely detailed study of the phenomena, we find that these two notions are idealizations, the probability of whose physical realization is nil.

* * *

Without continuing indefinitely this list of examples drawn from modern Physics, let us at once assume the philosophical point of view and ask the following question: May it not be universally true that the concepts produced by the human mind, when formulated in a slightly vague form, are roughly valid for Reality, but that when extreme precision is aimed at, they become ideal forms whose real content tends to vanish away? It seems to me that such is, in fact, the case, and that innumerable examples can be found in all spheres, particularly in those of Psychology and Ethics, as well as of everyday life.

Let us take an example from the ethical field and consider the concept of an honest man. Let us begin with a somewhat vague definition; let us say that an honest man is a man of great probity, who always tends to do what he considers his duty and to resist all temptations drawing him in the opposite direction. We shall find around us—for we must not be too pessimistic—a certain number of people who fulfil this definition. But if we were to insist that the crown of honesty is to be awarded only to a man who never, in any circumstances, at any moment of his life, experienced the slightest temptation to disobey his conscience, then no doubt we shall find a striking diminution—since human nature is full of frailty—in the number of men to whom our definition will apply. The more precise and rigid the concept becomes, that is to say, the more restricted becomes its sphere of application. Like the plane monochromatic wave, absolute virtue, if defined with too exacting a precision, is an idealization the probability of whose full realization tends to vanish away.

Examples of this kind are, it should be repeated, innumerable. In the psychological, ethical and social sphere an uncompromisingly rigid definition or argument often leads away from, rather than towards, Reality. It is true that the facts tend to assume a certain order within the framework supplied by our reason; but it is no more than a tendency, and the facts invariably overflow if the framework is too exactly defined.

Thus in the region of the inexact sciences of human conduct, the strictness of the definitions varies inversely as their applicability to the world of Reality. But now the question arises, whether we have any right to compare this fact with those encountered during the development of modern Physics. Admittedly we are dealing with nothing more than an analogy whose applicability must not be overstressed; yet I believe that it is less superficial than might at first be thought. Whenever we wish to describe facts, whether of a psychological or an ethical nature, or belonging to the sphere of the physical or natural sciences, we are inevitably dealing with a Reality which is always infinitely complex and full of an infinity of shades on the one hand, and on the other with our understanding, which forms concepts which are always more or less rigid and abstract. A confrontation between Reality and understanding is inevitable, and as far-reaching a reconciliation as possible, desirable. It is certain that our concepts are capable of adaptation to Reality to a considerable extent, provided that we allow them a certain margin of indeterminateness: if it were otherwise, no argument relating to real facts could be effected on the basis of any order of ideas. What is more doubtful is whether such a correspondence can be maintained to the end, if we insist on eliminating the margin of indeterminateness and on effecting extreme precision in our concepts. Even in the most exact of all the natural sciences, in Physics, the need for margins of indeterminateness has repeatedly become apparent—a fact which, it seems to us, is worthy of the attention of philosophers, since it may throw a new and illuminating

light on the way in which the idealizations formed by our reas
become adaptable to Reality.

What has been said also throws light on the parts played resp
tively by the spirit of geometry, and that of intuition, in the devel
ment of human knowledge. The spirit of geometry is need
for without it we could give no degree of precision to our id
and arguments; our knowledge, without it, would always rem
vague and merely qualitative. But there is a need also for
spirit of intuition; it is required to recall to us without ceas
that reality is too fluid and too rich to be contained in its entir
within the strict and abstract framework of our ideas. Th
are certainly familiar notions to all who have given any though
the progress and value of human knowledge; yet it seems to
that the development of Physics might suggest to such thinl
some new reflections in the sense which it has been the aim of tl
pages to indicate. ·

2

TO THE MEMORY OF ÉMILE MEYERSON

Early in 1933 I had decided to produce a series of booklets on the Philosophy of the Sciences. It seemed to me that there could be no better person whom one could ask to inaugurate this series than Émile Meyerson. Some years earlier I had been placed in touch with this distinguished philosopher through a common friend, so that I was aware of the interest with which he was following the development of contemporary Physics, and also of the desire he felt to put into exact terms his attitude to the novel philosophical problems raised by this development. Meyerson was good enough to write a fine and profound study, called *Reality and Determinism in Quantum Physics*, to form the first booklet in the new series on scientific philosophy, a study which was keenly appreciated and has often been quoted since; and he went further, and did me the great honour of requesting a Preface for this small volume. It was a slightly embarrassing honour for me, since it was not my part to present to the philosophical and scientific public a writer as well known and appreciated as Émile Meyerson. At the same time I was anxious to fall in with his wish, and I was also glad to have the opportunity to express the admiration I felt for his work and the respectful sympathy which I experienced for him personally; with the result that I wrote the brief Preface which is here reproduced.

* * *

"It seemed to us that, to open this series of booklets on the Philosophy of the Sciences, no better qualified personality could be found than Émile Meyerson. All philosophers and many scientists are acquainted with the brilliant studies through which, for 25 years, Meyerson has been endeavouring, with an admirable consistency based on a vast knowledge of the History of Science, to ascertain the manner in which human reason proceeds when it tries to understand. According to him, then, human reason believes itself to have reached understanding, both in scientific research and in everyday life, only if it succeeds in carving out identities and permanencies from the fluid Reality of the physical world. In this way the common structure of all those physical theories which aim at grouping classes of phenomena by means of a network of equalities, of equations, is explained. These theories have the constant aim of eliminating as far as possible all diversity and real change, and of showing that the posterior is in some way contained within the antecedent. The complete realization of the ideal pursued by reason, if formulated in this way, must appear chimerical, since it would consist in merging the entire qualitative diversity and the entire progressive variations of the physical Universe within one absolute identity and permanence. But if this complete realization is impossible, the nature of the physical world nevertheless permits a partial success of these attempts at rationalization. For in the physical world there are objects which persist practically unchanged through time, and there are also classes of objects which resemble each other sufficiently to allow us to identify them by uniting them under a common concept. These are the 'fibres' of Reality, as Meyerson calls them, and our reason lays hold of them in the experience of everyday life and uses them to build up our habitual view of the external world. It is these 'fibres,' as well as other more subtle ones revealed to our knowledge by the delicate methods of experimental research, which the scientist's reason makes use of in its attempt to extract from the variety and flux of Reality the element of identity and permanence which it contains. Thanks to the existence of these 'fibres' some sciences are possible, though the strict ideal of Science cannot be realized. Here is the miracle: and this is the position expressed by Paul Valéry in words no doubt inspired by a reading of Meyerson's works: 'The human spirit is foolish because it seeks; it is great because it finds.'

"Ultimately, however, the Universe cannot be reduced to a vast tautology, and inevitably our scientific description of Nature must encounter 'irrational' elements which resist our attempts at identification. It is the unwearying effort of human reason to limit the number and reduce the sphere of these elements.

"I do not pretend that these few lines sum up the rich and profound thoughts of the author of *Identity and Reality*.[1] My aim has been simply to throw light on certain aspects which struck me on reading his works, and in the course of conversations which I have had the pleasure of holding with him during recent years, to the great advantage of my own personal knowledge.

"In complete opposition to all anti-intellectualist doctrines, Meyerson's *Critique* sees in the work done by Science an effort of an inestimable value; it considers this effort as one of the finest tasks of humanity, and insists that we owe it the fullest admiration and the most complete respect. But it also leads him to reject the narrow dogmatism which pretends to rest upon the present or future results of Science; for although scientific knowledge can advance indefinitely, it must by its essential character be limited and partial on account of the very goal which reason sets it: for, strictly, the goal is unattainable.

"By means of an intellectual activity triumphing over every obstacle Émile Meyerson has kept abreast of the progress of contemporary Physics in every field, and of the difficult problems which this progress has raised."

* * *

Such was the Preface I had drafted. Meyerson was able to see this, and also to read the proofs of the booklet containing his impressive work. But, unhappily, when the small volume was published he was no longer with us. After a long and painful illness, death had removed him from the affection and respect of his friends and admirers.

It is no easy matter to express in a few lines the loss which

[1] Published in "The Library of Philosophy" (George Allen and Unwin, Ltd.).

his death has caused to Philosophy and to the History of the various sciences. The Philosophy of the sciences had been enriched by a series of works, which will remain classics, upon the general theory of the advancement and course of scientific thought. Meyerson's central idea has always seemed to be that human reason does not think that it has really understood a fact unless it succeeds in showing how the fact was already contained implicitly in our previous knowledge and in identifying it, in a manner, with what is already given. Hence the importance for him of discovering permanencies in every branch of Science, and hence too the essential part played by the Principles of Conservation in physical and chemical theory. But while demonstrating this instinctive tendency of our reason, this important philosopher boldly drew attention to its paradoxical character, since it is the case that all the efforts towards identification made by reason would only end, if they were to succeed, in abolishing all diversity and heterogeneity—in other words, in a kind of negation of the very world which they aim at explaining. And if reason nevertheless succeeds in escaping from the vicious circle, and in erecting a Science which does indisputably advance, the explanation (according to Meyerson) is that it allows certain irrational elements to find their way into our theoretical constructions, it being precisely these elements whose more or less surreptitious introduction saves the totality of our successive identifications from being nothing more than a vast tautology. Like every philosophical doctrine of a like scope, that of Meyerson has been and will be the subject of debate. Perhaps the study of the revolutionary changes in Quantum Physics, which he had begun to study deeply, despite age and illness, with an admirable ardour, would have led him to modify certain aspects of his views, had he been left the time. Yet it is certain that his suggestions, based as they are on a firm foundation of proof, contain an important element of truth, and that modern Epistemology cannot today afford to disregard them. Meyerson's work will remain among the most important produced by the contemporary Philosophy of Science.

Of the history of the various sciences Meyerson possessed a deep knowledge such as is rare in these days, when the many brilliant advances of modern Science tend to withdraw attention from the study of earlier work and of stages which have been left behind. The pleasure of a long conversation with him was needed before his vast erudition could be appreciated. To listen to him was to learn with every word. He was acquainted with every part of the history of scientific thought; he knew the contributions and the experiments, the successes and the blunders of antiquity, of the Middle Ages, of the Renaissance and of modern and contemporary times. His conversation was a living lesson on the successive efforts of the human mind; and if the occasion arose he did not think it unfitting to season such instruction with some striking anecdote drawn from the vast store of his memory. To understand how closely Philosophy and the History of Science were linked for Meyerson, we must recall the method on which he invariably based his researches. On his view, a risk of grave error was involved in any attempt to study *a priori*, or even by introspection alone, the advance and methods of reason. The correct method for this investigation, he rightly thought, consisted in observing the progress of scientific thought as it actually took place by carefully studying its historical development, its hard beginnings, its mistakes, its gropings and its lasting, or even provisional, successes. For this reason the History of Science played a fundamental part both in his thought and in his books. All readers who have thought about his work have noticed how consistently the explanation of his own ideas is mingled with considerations of an historic or psychological kind; an intermixture which, if sometimes startling, is never wearisome. The arguments in support of his theories are drawn from the rudimentary thought of primitive peoples, or of those which have been left behind in the course of evolution. The scrutiny of scientific theories which ended in failure interested him no less than that of those which succeeded; for in both he found the same tendencies which are characteristic of the inmost functioning of human reason. He was sometimes fond of rehabilitating certain tentative theories which had been abandoned, showing that in their own day they were perfectly tenable, as in the case of the famous phlogiston theory, to which he has devoted some extremely interesting pages.

Émile Meyerson was great as a philosopher and as a scholar, but he was also a man of agreeable and kindly conversation, who could charm from the first moment, as all who had the privilege of meeting him can testify. While the name of Meyerson is destined to remain among those of the most eminent thinkers of our time, the memory of a man who was as amiable as he was eminent will remain dear to all those who knew him.

3

MACHINE AND SPIRIT [1]

The advancement of applied science is the result of the joint work of pure science and of technical progress. It has never been the primary aim of pure science to build machines, or to obtain results having an immediate practical utility; its ideal has always been a higher one: to satisfy one of the noblest urges of the human mind, it has always pursued a disinterested knowledge, a knowledge *per se*. But in the gradual process of reading the riddle of the Laws of Nature, it has also increased our power over the material world, with the result that many aspects of applied mechanics owe their existence to the disinterested labours of scientists. On the other hand, while applied science has an immediately utilitarian aim, and must be ranked below pure science from the purely intellectual point of view, it is none the less a form of intellectual activity which seeks by ingenious apparatus or delicate adjustments to obtain certain definite results, or to overcome certain definite difficulties. The machine—the offspring of pure and of applied science—is thus also the offspring of intelligence; and if there is any reason why we should admire our modern material civilization apart from the advantages which it brings for everyday life—advantages which, it must be added, have counterbalancing disadvantages—this its derivation is quite certainly that reason.

But the question arises whether the machine, the daughter of intelligence, by mastering our whole civilization and weighing so heavily on our existence, may not turn against the spirit whence it arose and crush it. Wholly absorbed in the material preoccupations of an existence which daily grows more hurried and complicated, is it not likely that humanity will turn away from meditation, and

[1] Part of an Address delivered at the prize-giving at the Lycée Pasteur, July 13, 1932.

will abandon pure thought and all those lofty forms of intellectual activity which are not only an honour to the human race, but also the essential condition of its progressive evolution? This, I think, is the first serious problem imposed by the intensive development of the mechanistic spirit, and, more generally, of human civilization under its material form.

I may say at once that when faced by this problem I am not disturbed; for I feel that today as much as, and perhaps even more than, at any other period human thought has demonstrated its power and boldness. I have in mind particularly—a thing which will cause you no surprise—the splendid growth of scientific, and particularly of physical theories which characterized the first thirty years of this twentieth century. You will allow me to say a few words on this, my own, particular subject.

The special characteristic of the important new theories of contemporary Physics, both the Theory of Relativity and the Quantum Theory, is a stupendous effort to enlarge the limits of thought, and to shake off certain *a priori* concepts which certain natural phenomena refused to fit. The Theory of Relativity, in the first place, has again set a question mark against the old concepts—concepts so aptly fitting immediate intuition—of absolute Space, Length and Time. It has shown that ultimate Reality consists in a kind of intimate union between Space and Time, and that the way in which we carve out our time and our space from this baffling Reality has no more than a relative validity, and depends on the reference system to which we happen to be attached. Even more boldly, the Quantum Theory gives up the notion of the continuity of physical phenomena in its attempt to interpret phenomena which in appearance are inexplicable, and seeks a solution of the riddles of the atomic world by approaches of a strange novelty. Whatever may be the future fate of these theories, they are in any case splendid intellectual efforts. They are efforts which have not only brought new life to Physics, but have also brought fresh food to philosophic thought by asking some of the great traditional questions in an unexpected form: for example, by raising again the entire problem of Causality and Individuality.

This vitality of contemporary thought, which is by no means confined to theoretical Physics and the Philosophy of the various sciences, but is found also in many other fields, allows us to claim that the mechanical spirit has done no harm to the intellectual activity of mankind, and that so far no decline in this activity is to be observed. Not only is it the case that the mechanical spirit has had no baneful results so far; we may go further, I think, and say that it has substantially fostered this progress. For one thing, it would be easy to show the beneficent part played in that respect by all the inventions which, since that of printing, have aided the spreading of thought, the speed of communications, the intensity of the exchange of ideas between individuals and nations: a theme so easy to work out, and also so commonplace, that you will allow me not to stress it. But there is another and less obvious point to which I should like, still simply as a scientist, to draw your attention. There is one special form of the mechanical art in which the machine becomes the servant of intellectual curiosity; this form is experimental technique—the technique which supplies the scientist with the necessary instruments for studying Nature and discovering its Laws. Every important step forward made by Astronomy, Physics, Chemistry or Biology had one essential condition—the previous existence or invention of certain apparatus; and as the sciences sought to extend their advance, so it became necessary for instrumental technique to develop and to expand in its delicate adjustments. Ultimately, indeed, it was the indications furnished by experiment that gave rise to the great movements of thought which I mentioned a moment ago. For Einstein would never have had the idea of creating the Theory of Relativity unless experiments of an extremely high degree of precision—i.e. requiring extremely delicate apparatus—had shown that it was impossible to demonstrate in a laboratory the absolute motion of the Earth in Space. Physicists, again, would never have thought of forming those concepts, so wonderful in their strangeness, to which the Quantum Theory leads, unless experiment had shown those discontinuities and dualities which disagree so completely with all the old ideas. Left to itself, theoretical Science would always tend to rest on its laurels; but experiment, by becoming continually more exact and delicate, has shown us more clearly each day that "there are more things in heaven and earth than are dreamt of in your philosophy." By pointing to the infinite complexity of Reality, experiment has broken the circle within which speculative thought might easily risk confining itself, if left to its own devices. And since experiments depend on the perfection of experimental technique, the machine today is in a sense one of the essentials of intellectual progress.

Hitherto I have been speaking of pure thought, whether scientific or philosophical. Now though it is true that the mechanistic spirit has done no harm to pure thought, and though the reverse seems actually to be the case, are there not other lofty forms of our inner life on which it has inflicted, and will inflict, harm? Will it not be a harmful influence on esthetic feeling and on art in its various forms? By the changes which it brings about in the whole of our existence, will it not react dangerously upon our moral life? These are difficult questions, and I do not feel that I am particularly well equipped to discuss them. Others will be able to judge better than I whether literature, poetry and the arts in general are in a state of decadence, or whether on the contrary they have drawn fresh life from new ways of thinking and a new technique and are about to enter new and hitherto unexplored roads. Others will be able to speak with greater authority, too, on the evolution of the moral conscience in contemporary mankind, and to tell you whether in their opinion it is becoming keener or blunt. At the same time, if I were allowed to express an opinion on matters so far outside my proper sphere, I would tell you that here again I am rather on the side of optimism. For the evolution of esthetic, as also of ethical, feeling has its roots in certain major tendencies of the human soul, and these tendencies are in turn linked with the deepest and most mysterious forces of life. They have followed mankind across the ages, and I do not think that they will fail before mankind itself ceases to be. It is true that bad tendencies exist by the side of the good. Our love of the beautiful is thwarted by bad taste and by the inadequacy of the effort to realize the ideal; our aspirations after the good are baulked by bad propensities, by egoism and indifference. But such has always been the case, and it has not prevented art and literature from flourishing, nor the sense of duty remaining ineradicable in the human heart. It is abundantly clear that the material circumstances of our life change and become more complicated; but despite all these changes we see humanity remaining the same; and the stability of its moral capital seems to assure the permanence of its aspirations.

Accordingly I do not think that modern civilization is bound to crush either thought, or sensibility, or conscience. But there are other aspects of the question which are obviously less reassuring. There are the harmful reactions caused by the excessive rate at which the machine has been developed, and by the over-intensity of production: the alternating attacks of fever and depression to which the economic organism is liable, and which exhaust the patient. More dangerous still is the growing power of destruction, the deplorable if inevitable consequence of the development of applied science. It has been said that knowledge is power; but unhappily it is power for evil as well as for good. Modern humanity, with the means of destruction which lie ready to its hand, may be compared to a child holding a loaded revolver and ignorant of the danger to which it is exposed while handling it. Let us hope that the child will reach the age of responsibility before an irremediable catastrophe occurs. Yet one cannot but feel some anxiety.

Here then is matter of serious reflection for the thoughtful spirits of our time; and there are other subjects which I cannot examine here. Yet it is wise not to take too dark a view and to avoid lapsing into pessimism. Since the beginnings of its history, the human race has encountered many obstacles and has met many dangers: at the cost of great efforts and, it must be admitted, also of great suffering, it has, on the whole, overcome them; the best proof of this is that we are here at this moment. Thus it may reasonably be permitted to hope that the case will be the same in the future, and that we shall move along the edge of the precipice without falling.

The danger inherent in too intense a material civilization, to sum up, is not that civilization itself: it is the disequilibrium which would result if a parallel development of the spiritual life were to fail in providing the needful balance. Here is the principal cause of the difficult but essential part played by the training and education of the younger generation, whose duty it will be to learn how to profit by the practical advantages of modern life, without at the same time ceasing to preserve and even augment the moral heritage which mankind has slowly accumulated in the course of the centuries. To be able to do this, however, the younger generations must undoubtedly acquire learning, so that they shall be able to benefit by past experience and by the stock of gathered knowledge; but, beyond learning, they must acquire a love of personal endeavour, the love of beauty in all its forms, of the art of thinking and of the expression of thought. By the education which you are receiving today you are being fitted to take your place among the *élite* of the nation tomorrow. On you the fate of our civilization will, then, in part depend. Let me conclude by hoping that you will preserve throughout your life, as a precious fruit of your work here, the zeal to pursue all that is loftiest in the intellectual, the esthetic or the moral sphere. For without such zeal a civilization, however great the perfection of its material aspects, would soon be no better than a complicated form of barbarism.

INDEX